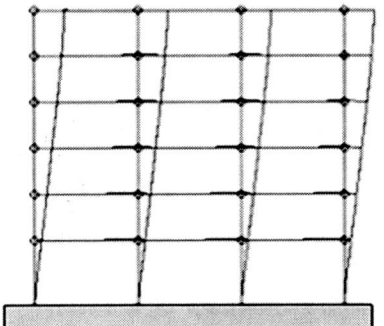

Seismic Loads

Other Titles of Interest

Minimum Design Loads for Buildings and Other Structures, ASCE/SEI 7-05. (ASCE Standard, 2006). Provides requirements for general structural design and includes means for determining dead, live, soil, flood, wind, snow, rain, atmospheric ice, and earthquake loads and their combinations that are suitable for inclusion in building codes and other documents. A detailed commentary of explanatory and supplementary information is included. (ISBN 978-0-7844-0809-4)

Books Related to ASCE 7-05

Snow Loads: Guide to the Snow Load Provisions of ASCE 7-05, **by Michael O'Rourke, Ph.D., P.E.** (ASCE Press, 2007). Presents a detailed, authoritative interpretation of the snow load provisions of ASCE 7-05 by a respected engineering professional, plus three complete design examples. (ISBN 978-0-7844-0857-5)

Wind Loads: Guide to the Wind Load Provisions of ASCE 7-05, **by Kishor C. Mehta, Ph.D., P.E., and William Coulbourne, P.E.** (ASCE Press, 2010). Presents a detailed, authoritative interpretation of the wind load provisions of ASCE 7-05 by respected engineering professionals. (ISBN 978-0-7844-0858-2)

Books on Seismic Engineering

Earthquake-Actuated Automatic Gas Shutoff Devices, ANSI/ASCE/SEI 25-06. (ASCE Standard, 2008). Provides current minimum functionality requirements for earthquake-actuated automatic gas shut-off devices and systems. (ISBN 978-0-7844-0877-3)

Seismic Rehabilitation of Existing Buildings, ASCE/SEI 41-06. (ASCE Standard, 2007). Describes the latest generation of performance-based seismic rehabilitation methodologies in a new national consensus standard. (ISBN 978-0-7844-0884-1)

Seismic Evaluation of Existing Buildings, ASCE/SEI 31-03. (ASCE Standard, 2002). Instructs design professionals, code officials, and building owners on how to evaluate whether existing buildings are adequately designed and constructed to resist seismic forces (ISBN 978-0-7844-0670-0)

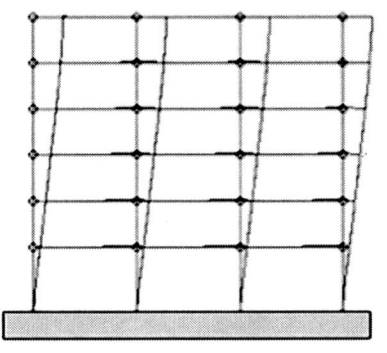

Seismic Loads

Guide to the Seismic Load Provisions of ASCE 7-05

Finley A. Charney, Ph.D., P.E.

PRESS

Library of Congress Cataloging-in-Publication Data
Charney, Finley Allan.
  Seismic loads : guide to the seismic load provisions of ASCE 7-05 / Finley A. Charney.
    p. cm.
  Includes bibliographical references and index.
  ISBN 978-0-7844-1076-9
  1. Earthquake resistant design–Standards. 2. Earthquake resistant design–Case studies. I. American Society of Civil Engineers. II. Title.
  TA658.44.C38 2010
  624.1'762–dc22
                                        2009048315

Published by American Society of Civil Engineers
1801 Alexander Bell Drive
Reston, Virginia 20191
www.pubs.asce.org

Any statements expressed in these materials are those of the individual authors and do not necessarily represent the views of ASCE, which takes no responsibility for any statement made herein. No reference made in this publication to any specific method, product, process, or service constitutes or implies an endorsement, recommendation, or warranty thereof by ASCE. The materials are for general information only and do not represent a standard of ASCE, nor are they intended as a reference in purchase specifications, contracts, regulations, statutes, or any other legal document.

ASCE makes no representation or warranty of any kind, whether express or implied, concerning the accuracy, completeness, suitability, or utility of any information, apparatus, product, or process discussed in this publication, and assumes no liability therefor. This information should not be used without first securing competent advice with respect to its suitability for any general or specific application. Anyone utilizing this information assumes all liability arising from such use, including but not limited to infringement of any patent or patents.

ASCE and American Society of Civil Engineers—Registered in U.S. Patent and Trademark Office.

Photocopies and reprints. You can obtain instant permission to photocopy ASCE publications by using ASCE's online permission service (http://pubs.asce.org/permissions/requests/). Requests for 100 copies or more should be submitted to the Reprints Department, Publications Division, ASCE (address above); e-mail: permissions@asce.org. A reprint order form can be found at http://pubs.asce.org/support/reprints/.

Copyright © 2010 by the American Society of Civil Engineers.
All Rights Reserved.

ISBN 978-0-7844-1076-9

Manufactured in the United States of America.
18 17 16 15 14 13 12 11 10 1 2 3 4 5

Contents

Preface vii
Abbreviations and Symbols xi
Table of Conversion Factors xiii

Examples
    1. Occupancy Category 1
    2. Importance Factor and Seismic Design Category 7
    3. Site Classification Procedure for Seismic Design 11
    4. Determining Ground Motion Parameters 19
    5. Developing an Elastic Response Spectrum 25
    6. Ground Motion Scaling for Response History Analysis 29
    7. Selection of Structural Systems 41
    8. Combinations of Lateral Load Resisting Systems 49
    9. Horizontal Structural Irregularities 57
    10. Vertical Structural Irregularities 69
    11. Diaphragm Flexibility 81
    12. Structural Analysis Requirements 89
    13. Determining the Redundancy Factor 93
    14. Accidental Torsion and Amplification of Accidental Torsion ... 99
    15. Load Combinations 109
    16. Effective Seismic Weight (Mass) 119
    17. Period of Vibration 129
    18. Equivalent Lateral Force Analysis 139
    19. Drift and P-Delta Effects 153
    20. Modal Response Spectrum Analysis 165
    21. Diaphragm Forces 185

Frequently Asked Questions 189

Appendixes
    A. Interpolation Functions 207
    B. Using the USGS Seismic Hazards Mapping Utility 211
    C. Using the PEER NGA Database 217

References 221
Index 223
About the Author 233

Preface

The purpose of this guide is to provide the reader with a set of examples related to the use of ASCE Standard 7, *Minimum Design Loads for Buildings and Other Structures* (ASCE/SEI 7-05). The guide is also pertinent to users of the *2006 International Building Code* (ICC 2006) because IBC refers directly to ASCE 7.

Sections of ASCE 7 Pertinent to the Guide

The guide has examples pertinent to the following chapters of ASCE 7:

Chapter 1: General
Chapter 2: Combinations of Loads
Chapter 11: Seismic Design Criteria
Chapter 12: Seismic Design Requirements for Building Structures
Chapter 16: Seismic Response History Procedures
Chapter 20: Site Classification Procedure for Seismic Design
Chapter 22: Seismic Ground Motion and Long Period Transition Maps

Seismic material excluded from the guide are Chapters 13 (Seismic Design Requirements for Nonstructural Components), Chapter 14 (Material-Specific Design and Detailing Requirements), Chapter 15 (Seismic Design Requirements for Nonbuilding Structures), Chapter 17 (Seismic Design Requirements for Seismically Isolated Structures), Chapter 18 (Seismic Design Requirements for Structures with Damping Systems), Chapter 19 (Soil Structure Interaction for Seismic Design), and Chapter 21 (Site-Specific Ground Motion Procedures for Seismic Design).

The vast majority of the examples in the guide relate to Chapters 1, 2, 11, and 12 of ASCE 7; the principal subject covered is buildings. The materials on nonstructural components and on nonbuilding structures will be expanded in a later edition of this book or in a separate volume. The materials presented for Chapter 16 relate only to the selection and scaling of ground motions for response history analysis.

Chapter 14 of ASCE 7 is not included because the guide is focused principally on seismic load analysis, and not seismic design. The reader is referred to the reference section of the guide for resources containing design examples. The materials included in Chapters 17 through 19 are considered advanced topics and may be included in a separate volume of examples.

The principal purpose of this guide is to illustrate the provisions of ASCE 7, and not to provide background on the theoretical basis of the provisions. Hence, theoretical discussion is held to a minimum. However, explanations are provided in a few instances. The reference section of this guide contains several sources for understanding the theoretical basis of the ASCE 7 seismic loading provisions. Specifically, the reader is referred to the commentary section to the *NEHRP Recommended Provisions for Seismic Regulations for New Buildings and Other Structures*. Also of interest to the reader is FEMA 451CD, the *NEHRP Recommended Provisions: Design Examples*.

How to Use the Guide

This guide is organized into a series of individual examples. With minor exceptions, each example stands alone and does not depend on information provided in other examples. This arrangement means that, in some cases, information is provided in the beginning of the example that requires some substantial calculations, but these calculations are not shown. For instance, in the example on drift and P-delta effects, the details for computing the lateral forces used in the analysis are not provided, and insufficient information is provided for the reader to backcalculate these forces. However, reference is made to other examples in the guide where similar calculations (e.g., finding lateral forces) are presented. The reader should always be able to follow and reproduce all new numbers (not part of the given information) that are generated in the example.

Table and Figure Numbering

The examples presented in the guide often refer to sections, equations, tables, and figures in ASCE 7. All such items are referred to directly, without specific reference to ASCE 7. For instance, a specific example might contain the statement "The response modification factor R for the system is provided by Table 12.2-1."

References to sections, equations, tables, and figures that are unique to the guide are always preceded by the letter G and appear in boldface. For example, the text may state that the distribution of forces along the height of the structure is listed in **Table G12-3** and illustrated in **Figure G12-5**. In this citation, the number 12 is the example number, and the number following the dash is the sequence number of the item (i.e., third table, fifth figure).

Notation and Definitions

The mathematical notation in the guide follows directly the notation provided in Chapter 11 of ASCE 7. A list of Abbreviations and Symbols provides definitions for symbols that have been introduced in the guide.

Computational Units

All of the examples in the guide are developed in the U.S. Customary (English) system, as follows (with the standard abbreviation in parentheses):

Length units:	inches (in.) or feet (ft)
Force units:	pounds (lb) or kips
Time units:	seconds (s)

All other units (e.g., mass) are formed as combinations of the above units.
A table with common unit conversions is provided.

Appendixes

The three appendixes provide additional detail that is not necessary in specific examples. Appendix A covers interpolation functions, Appendix B describes how to use the U.S. Geological Survey Seismic Hazards Mapping Utility, and Appendix C explains the use of the PEER Web site, from which ground motion acceleration histories may be obtained.

User Comments

The author invites users to comment on any ambiguities or errors that are found in this guide. Suggestions for improvement or additions are welcomed and will be considered for future versions of the guide, especially the one to accompany the forthcoming ASCE/SEI 7-10.

Disclaimer

The interpretations of ASCE 7 requirements, as well as all other opinions presented in this guide, are those of the author and do not necessarily represent the views of the ASCE 7 Standards Committee or the American Society of Civil Engineers.

Acknowledgments

The author wishes to acknowledge the following individuals for their contributions to this book:

James R. Harris, P.E., Ph.D. (advisor)

John Hooper, P.E., Magnusson Klemencic, Seattle, Washington (reviewer)

William P. Jacobs V, P.E., Stanley D. Lindsey & Associates, Ltd., Atlanta, Georgia (reviewer)

Viral Patel, P.E., Walter P. Moore and Associates, Austin, Texas (reviewer)

Charles J. Smith, P.E., and Thomas T. Moore, E.I.T., Schnabel Engineering Associates, Blacksburg, Virginia (author of the example on Site Classification Procedure)

Abbreviations and Symbols

Abbreviations

2D	Two-dimensional
3D	Three-dimensional
ACI	American Concrete Institute
AISC	American Institute of Steel Construction
ASCE	American Society of Civil Engineers
ASTM	Formerly American Society for Testing and Materials, now ASTM International
BRB	Buckling restrained brace
CBF	Concentrically braced frame
CQC	Complete quadratic combination
DBE	Design basis earthquake
EBF	Eccentrically braced frame
ELF	Equivalent lateral force
FEMA	Federal Emergency Management Agency
IBC	International Building Code
LRH	Liner response history
MCE	Maximum considered earthquake
MRS	Modal response spectrum
NGA	Next generation attenuation
NRH	Nonlinear response history
OC	Occupancy Category
PEER	Pacific Earthquake Engineering Research Center
RC	Reinforced concrete
SDC	Seismic Design Category
SRSS	Square root of the sum of squares

Symbols Unique to This Guide

Symbol	Definition	Introduced in Example No.
CS	Combined scale factor	6
FPS	Fundamental period scale factor	6
k	Lateral stiffness of component	9
K	Structural stiffness matrix	20
M	Structural mass matrix	20
M_{pCk}	Plastic moment strength of column k	10
M_{pGk}	Plastic moment strength of girder k	10
R	Modal excitation vector	20
R_{eff}	Effective response modification coefficient	7
S_{ai}	Spectral acceleration in mode i	20
S_{di}	Spectral displacement in mode i	20
$T_{computed}$	Period computed by structural analysis	17
T_{MF}	Period at which Eq. 12.8-5 controls C_s	7
T_{MN}	Period at which Eq. 12.8-6 controls C_s	18
V_{yi}	Story strength	10
G	Modal participation factor	20
δ_i	Displacement in mode i	20
Δ_o	Drift computed without P-delta effects	19
Δ_{center}	Drift at geometric center of building	9
Δ_{edge}	Drift at edge of building	9
Δ_F	Drift computed with P-delta effects	19
ϕ	Mode shape	20
w	Circular frequency of vibration	20

Table of Conversion Factors

U.S. customary units	International System of Units (SI)
1 inch (in.)	25.4 millimeters (mm)
1 foot (ft)	0.3048 meter (m)
1 statute mile	1.6093 kilometers (km)
1 square foot (ft^2)	0.0929 square meter (m^2)
1 cubic foot (ft^3)	0.0283 cubic meter (m^3)
1 pound (lb)	0.4536 kilogram (kg)
1 pound (force)	4.4482 newtons (N)
1 pound per square foot (lb/ft^2)	0.0479 kilonewton per square meter (kN/m^2)
1 pound per cubic foot (lb/ft^3)	16.0185 kilograms per cubic meter (kg/m^3)
1 degree Fahrenheit (°F)	1.8 degrees Celsius (°C)
1 British thermal unit (Btu)	1.0551 kilojoules (kJ)
1 degree Fahrenheit per British thermal unit (°F/Btu)	1.7061 degrees Celsius per kilojoule (°C/kJ)

Example 1
Occupancy Category

This example demonstrates the selection of Occupancy Category for a variety of buildings and other structures.

Occupancy Category is used in several places in ASCE 7, including

- determination of importance factor (Section 11.5.1 and Table 11.5-1)
- requirements for protected access for Occupancy Category IV structures (Section 11.5.2)
- determination of Seismic Design Category (Section 11.6 and Tables 11.6-1 and 11.6-2)
- determination of drift limits (Section 12.12.1 and Table 12.12-1).

Table 1-1 provides guidance for determining the Occupancy Category. The Occupancy Categories range from I (nonessential facility with relatively low consequence of operational inhibiting damage) to IV (essential facility with great consequence of operational inhibiting damage).

It is important to note that when ASCE 7 is used in association with the *2006 International Building Code* (ICC 2006; referred to here as IBC), Table 1604.5 of IBC supersedes Table 1-1 of ASCE 7. Both tables provide four occupancy categories (I through IV), and for the most part, the definitions are similar. There are differences, however. For example, Table 1604.5 of IBC lists "Any other occupancy with an occupancy load greater than 5000" as an Occupancy Category III structure, whereas ASCE 7 has no such entry, and hence, if ASCE 7 were used alone, the same building would be assigned to Occupancy Category II.

In the following examples, the use of Table 1-1 of ASCE 7 and Table 1604.5 of IBC is demonstrated through several scenarios. For each example, the structure is briefly described, and the Occupancy Category is presented followed by discussion. For most of the examples, the same classification arises from ASCE 7 and IBC. Brief discussion is provided when there are differences.

Note that selection of Occupancy Category can be somewhat subjective. When in doubt, the local building official should be consulted.

While each individual example in this chapter provides a geographic location for the building or structure under consideration, this location is not relevant to the selection of the Occupancy Category. These locations are provided simply to add some realism to the scenarios given.

Exercise 1 A three-story university office and classroom building in Blacksburg, Virginia. Highest occupancy per classroom = 60 students. Maximum building capacity = 375.

Answer Occupancy Category = II.

Explanation An Occupancy Category of II was chosen because the building has a total capacity of fewer than 500 individuals, and no more than 300 people congregate in one room. Note that a high school (secondary school) building with an identical configuration would have an Occupancy Category of III.

Exercise 2 A six-story medical office building with outpatient surgical facilities located in Austin, Texas.

Answer Occupancy Category = IV.

Explanation An Occupancy Category of IV is used here because the building has surgical facilities. This category is used even though the building has no accommodation for extended or overnight patient care. It is possible that the Occupancy Category could be reduced to II if the surgical component of the facility is deemed to be insignificant in the context of emergency response.

Exercise 3 A one-story elderly care facility (Alzheimer's care and nursing home) with a maximum of 120 residents located in Savannah, Georgia.

Answer Occupancy Category = III.

Explanation The Occupancy Category is III because the facility has no surgery or emergency care capability.

Exercise 4 A 40-story casino and hotel in Reno, Nevada. Gambling rooms, ballrooms, and theaters accommodate as many as 800 people each. Total hotel capacity is more than 10,000 people.

Answer Occupancy Category = III (ASCE 7).
Occupancy Category = III (IBC).

Explanation The Occupancy Category of III is selected from ASCE 7 Table 1-1 because the structure has several rooms in which more than 300 people may congregate.
For IBC, an Occupancy Category of III is selected because the occupancy load is greater than 5,000.

Exercise 5 Municipal courthouse and office building, containing two prisoner holding cells (a maximum of 15 prisoners in each) and a sheriff's department radio dispatcher facility, located in Richmond, California. Courtrooms have a maximum capacity of 120.

Answer Occupancy Category = IV.

Explanation Here the driving factor is the sheriff's department radio dispatcher facility, which is necessary for post-earthquake communication. Had this radio facility not been present, the Occupancy Category would reduce to II or III, depending on how the prisoner holding cells were classified.

Exercise 6 Retail fireworks building in Chattanooga, Tennessee, approximately 10,000 ft^2.

Answer Occupancy Category = II.

Explanation Although fireworks can be considered explosive, the energy released by the explosions is relatively small (compared to, for example, a facility that stores military munitions). For this reason, the Occupancy Category of II is selected. This is a case where a discussion with the local building official would be useful because there is some potential for classification as Occupancy Category III.

Exercise 7 CBS News Affiliates office building in Tallahassee, Florida, that contains two studios for broadcasting local news. The facility is not a designated emergency communication center.

Answer Occupancy Category = III (ASCE 7).
Occupancy Category = II (IBC).

Explanation The Occupancy Category of III for ASCE 7 is based on the classification of the building as a telecommunication center. Occupancy Category IV is not appropriate because this is not a designated emergency communication center.
   An Occupancy Category of II is selected for IBC because IBC does not list telecommunication centers in its description of Occupancy Category III uses.

Exercise 8 A 95-story, mixed-use building in Chicago, Illinois, containing two floors of retail facilities (shops and restaurants with a maximum capacity of 60), 50 levels of office building space, and 43 levels of apartments.

Answer Occupancy Category = II (ASCE 7).
Occupancy Category = II or III (IBC).

Explanation This building could have very high occupancy (several thousand people) at any time, but large numbers of people do not congregate in any one area. For this reason, the Occupancy Category for ASCE 7 is II. For IBC, a building with an occupancy load greater than 5,000 is assigned an Occupancy Category of III. It is possible that a building of this size could have an occupancy load greater than 5,000.

Exercise 9 One-story Greyhound bus station in Santa Fe, New Mexico. Buses enter the facility to load and unload passengers. Maximum occupancy of structure = 350.

Answer Occupancy Category = III.

Explanation The Occupancy Category of III is selected from Table 1-1 of ASCE 7 because of the maximum occupancy of 350 people. An Occupancy Category of III is used for IBC because the primary use of the facility (awaiting transportation) is classified as public assembly.

Exercise 10	Beer manufacturing, warehouse, and distribution facility in Golden, Colorado.
Answer	Occupancy Category = II.
Explanation	The Occupancy Category of II is used because higher categories are not appropriate. The facility cannot be designated as Occupancy Category I because it is not a minor storage facility.
Exercise 11	Grandstand for a college football stadium with seating for 15,000 individuals in Lubbock, Texas.
Answer	Occupancy Category = III.
Explanation	For this structure, more than 300 people congregate in one area, so the Occupancy Category of III is required.
Exercise 12	Dockside cargo storage warehouse adjacent to the Houston Ship Channel. The building is one story with 30,000 ft^2. Cargo is transported by forklifts and overhead cranes and is moved in and out daily. The cargo may contain materials (certain liquids in metal drums) considered toxic to humans.
Answer	Occupancy Category = III (but reducible to II on the basis of Section 1.5.2).
Explanation	Although the principal use of this nonessential facility is for storage, there is significant human activity in the structure, so a classification of Occupancy Category I is not appropriate. The storage of toxic materials generally requires an Occupancy Category of III. According to Section 1.5.2 of ASCE 7, the structure may be classified as Occupancy Category II if it can be demonstrated through a hazard assessment and risk management plan that the release of toxic materials does not pose a threat to the public.
Exercise 13	Grain storage silo in Hays, Kansas.
Answer	Occupancy Category = I.
Explanation	This assignment is based on the classification of this nonbuilding structure as an agricultural facility.

Exercise 14 Pedestrian bridge between an NFL football stadium and an adjacent parking lot. One end of the bridge is supported by the stadium superstructure. The bridge spans over a spur of an interstate highway. Estimated maximum number of people on the bridge at any time = 220. The bridge is in San Antonio, Texas.

Answer Occupancy Category = III.

Explanation The stadium would have an Occupancy Category of III. The bridge is given an Occupancy Category of III because it is a means of egress from the stadium and is thereby classified as part of the stadium. (An additional consideration is the fact that a full or partial collapse of the bridge onto the interstate would inhibit the movement of emergency vehicles.)

Exercise 15 The entry foyer for a regional hospital in St. Louis, Missouri. The 1,800-ft^2, glass-enclosed structure is separate from but closely adjacent to the hospital. The main purpose of the foyer is for visitors to the hospital to gain access to the main hospital building. A reception desk and several unmanned information kiosks are also in the foyer. There is a covered walkway between the main hospital and the foyer. Hospital staff and emergency personnel gain access through other portals.

Answer Occupancy Category = II or IV.

Explanation The hospital is clearly an Occupancy Category IV facility. If the foyer was an operational entry into the hospital, Section 11.5.2 of ASCE 7 would require the foyer to be classified as Occupancy Category IV as well. However, because hospital staff and emergency personnel do not gain access through the foyer, the entry foyer may be considered nonoperational, and hence, may be classified as an Occupancy Category II structure. Consultation with the local building official would be appropriate before a final designation could be made.

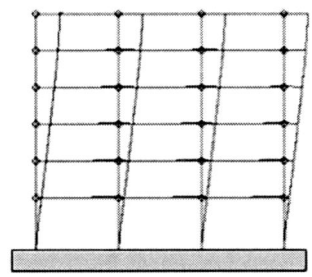

Example 2

Importance Factor and Seismic Design Category

This example demonstrates the determination of the importance factor and the Seismic Design Category.

2.1 Importance Factor

Importance factors are a function of only the Occupancy Category and are provided in Table 11.5-1. Values range from 1.0 for Occupancy Categories I and II to 1.5 for Occupancy Category IV. The primary use of the importance factor (I) is in the determination of design lateral forces. For example, Eq. 12.8-2 provides the response coefficient C_s for low period systems

$$C_s = \frac{S_{DS}}{\left(\dfrac{R}{I}\right)} \qquad (12.8\text{-}2)$$

which may be re-written as

$$C_s = I\frac{S_{DS}}{R}$$

From the revised equation, it is seen that I appears to act as a multiplier on the seismic strength demand. However, the factor should more realistically be interpreted as a multiplier on the required seismic capacity because the true demand (from the earthquake) is not a function of I. For example, the building designed for $I=1.5$ would have 1.5 times the lateral seismic strength of the same building designed for $I=1.0$, but both buildings are designed for the *same* ground motions.

I is not included (cancels out) in computed interstory drifts that are used in association with the drift limits given in Table 12.12-1. This fact can be seen from Eq. 12.8-15:

$$\delta_x = \frac{C_d \delta_{xe}}{I} \tag{12.8-15}$$

where the importance factor I is included in the lateral loads that produce δ_{xe}. On the other hand, the allowable drifts provided in Table 12.12-1 are a function of the Occupancy Category, and via Table 11.5-1, are also directly related to the importance factor. On this basis, an Occupancy Category IV building of a given R value, with $I=1.5$, would be designed to be 1.5 times stronger than, and for most structures (see "All other structures" in Table 12.12-1) would have one-half of the allowable drift of, the same building designed with an Occupancy Category of I or II.

The comparison of system behaviors with different Importance Factors is shown through a set of idealized force-deformation plots in **Fig. G2–1**. The Occupancy Category IV building with $I=1.5$ would have a significantly lower ductility demand, and probably less damage, than the system with $I=1.0$. Damage is reduced, but is not eliminated, in Occupancy Category IV systems.

2.2 Seismic Design Category

Seismic Design Category (SDC) is defined in Section 11.6 and by Tables 11.6-1 and 11.6-2. The parameters that affect SDC are the Occupancy Category and the design level spectral accelerations S_{DS} and S_{D1}, or for very high level ground motions, the mapped Maximum Considered Earthquake (MCE) (Section 11.4) spectral acceleration S_1. The SDC depends on the site class because S_{DS} and S_{D1} are directly related to the site class via Eqs. 11.4-1 and 11.4-2.

In the examples that follow, two sites, one in southwest Virginia and the other in Berkeley, California, are considered. For each site, consideration is given to the same structure constructed on soils with Site Class B or D. Consideration is given also to two different Occupancy Categories for each site: II and IV. The results of the calculations are presented in **Table G2–1** and **Table G2–2** for the southwest Virginia and Berkeley locations, respectively.

For the southwest Virginia site, which is of relatively low seismicity, the Seismic Design Category for the Site Class B location is B for the Occupancy Category II buildings and C for the Occupancy Category IV buildings. For Site Class D, the Seismic Design Category is C for the Occupancy Category II building and D for the Occupancy Category IV building. The elevation in Seismic Design Category on the Site Class D soils is caused by the site class amplification factors F_a and F_v (see calculations below). Thus, in southwest Virginia, the Seismic Design Category can range from B to D, depending on use and site. Moving from SDC B to D has important implications for the design and detailing of the structural system.

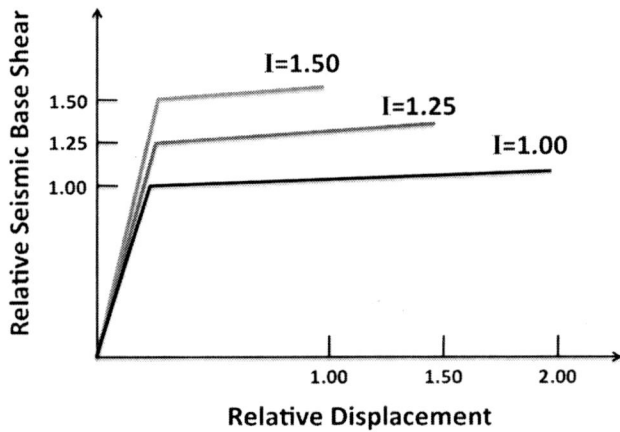

Figure G2-1 Influence of Importance Factor on System Performance for Systems Designated as "All Other Structures" in Table 12.12-1

Table G2-1 Determination of Seismic Design Category for Sites in Southwest Virginia

Site Class	Ground Motion Parameters (g)				Seismic Design Category	
	S_S	S_I	S_{DS}	S_{DI}	II	IV
B	0.35	0.10	0.233	0.067	B	C
D	0.35	0.10	0.355	0.160	C	D

Table G2-2 Determination of Seismic Design Category for Sites in Berkeley, California

Site Class	Ground Motion Parameters (g)				Seismic Design Category	
	S_S	S_I	S_{DS}	S_{DI}	II	IV
B	1.65	0.65	1.10	0.43	D	D
D	1.65	0.65	1.10	0.65	D	D

In Berkeley, the SDC is D in all cases. Had the ground motions been somewhat greater, with S_1 greater than 0.75 g, the Seismic Design Category would be elevated to SDC E for the Occupancy Category I, II, and III structures and elevated to SDC F for structures with Occupancy Category IV.

2.2.1 Detailed Calculations for Southwest Virginia

From Figs. 22-1 and 22-2, $S_S = 0.35$ g, and $S_1 = 0.10$ g.

For Site Class B

$F_a = 1.0$ and $F_v = 1.0$ (Tables 11.4-1 and 11.4-2)

$$S_{DS} = \frac{2}{3} F_a S_S = \frac{2}{3}(1.0)(0.35) = 0.233g$$ (Eqs. 11.4-1 and 11.4-3)

$$S_{D1} = \frac{2}{3} F_v S_1 = \frac{2}{3}(1.0)(0.10) = 0.067g$$ (Eqs. 11.4-2 and 11.4-4)

Seismic Design Category = B
for Occupancy Category II (Tables 11.6-1 and 11.6-2)
Seismic Design Category = C
for Occupancy Category IV (Tables 11.6-1 and 11.6-2)

For Site Class D

$F_a = 1.52$ and $F_v = 2.40$ (Tables 11.4-1 and 11.4-2)

$$S_{DS} = \frac{2}{3} F_a S_S = \frac{2}{3}(1.52)(0.35) = 0.355g$$ (Eqs. 11.4-1 and 11.4-3)

$$S_{D1} = \frac{2}{3} F_v S_1 = \frac{2}{3}(2.4)(0.10) = 0.160g$$ (Eqs. 11.4-2 and 11.4-4)

Seismic Design Category = C
for Occupancy Category II (Tables 11.6-1 and 11.6-2)
Seismic Design Category = D
for Occupancy Category IV (Tables 11.6-1 and 11.6-2)

Note: See Example 4 and Appendix A for determination of F_a and F_v.

2.2.2 Seismic Design Category Exception for Buildings with Short Periods

Under certain circumstances, it is permitted to determine the Seismic Design Category on the basis of S_{DS} only. The specific requirements are listed as four numbered points in Section 11.6. This provision is applicable only for systems with very short periods of vibration (with the approximate period T_a less than $0.8T_S$). This exception, where applicable, may result in the lowering of the SDC from, say, C to B, where the SDC of C would be required if the exception were not evaluated.

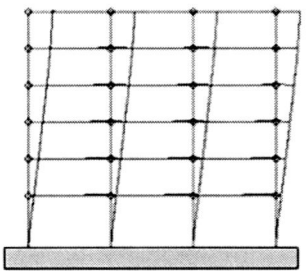

Example 3
Site Classification Procedure for Seismic Design

In this example, the seismic site class is determined for a given site.

Site class is used to characterize the type and properties of soils at a given site and account for their effect on the site coefficients, F_a and F_v, used in developing the design response spectrum (generalized simplified seismic analysis). The procedure can also require a site response analysis in accordance with Section 21.1, depending on the site class determination. However, the site classification procedure does not encompass evaluation of potential geologic and seismic hazards (Section 11.8). This example is applicable to the site classification procedure set forth in Chapter 20 of ASCE 7-05. See Chapter 11 for definitions pertaining to the site classification procedure.

Based on the competency of the soil and rock material, a site is categorized as Site Class A, B, C, D, E, or F. The site classes range from hard rock to soft soil profiles as presented in **Table G3–1**. This table appears in ASCE 7-05 as Table 20.3-1.

For this example, the shear wave velocity criteria are not covered in detail. Shear wave velocity correlations and direct measurement require considerable experience and judgment, which are beyond the scope of this

Table G3–1 Site Classification

Site Class	$\overline{V}_s$	$\overline{N}$ or $\overline{N}_{ch}$	$\overline{S}_u$
A. Hard rock	> 5,000 ft/s	NA	NA
B. Rock	2,500 to 5,000 ft/s	NA	NA
C. Very dense soil and soft rock	1,200 to 2,500 ft/s	> 50	> 2,000 lb/ft²
D. Stiff soil	600 to 1,200 ft/s	15 to 50	1,000 to 2,000 lb/ft²
E. Soft clay soil	< 600 ft/s	< 15	< 1,000 lb/ft²
	Any profile with more than 10 ft of soil having the following characteristics: - Plasticity index PI > 20, - Moisture content $w \geq 40\%$, and - Undrained shear strength u < 500 lb/ft²		
F. Soils requiring site response analysis in accordance with Section 21.1	See Section 20.3.1		

example. Proper use of shear wave velocity data requires consulting with an experienced professional.

3.1 Gathering Data

To classify a site, the proper subsurface profile and necessary data need to be obtained. According to Section 20.1:

- Site soil shall be classified based on the upper 100 ft (30 m) of the site profile.
- In the absence of data to a depth of 100 ft, soil properties are permitted to be estimated by the registered design professional preparing the soil investigation.
- Where soil properties are not known in sufficient detail, Site Class D shall be used unless Site Class E or F soils are determined to be present.

Where site class criteria are based on soil properties (PI, w, s_u), the values are to be determined by laboratory tests as specified: Atterberg Limits (ASTM D4318), moisture content (ASTM D2216), and undrained shear strength (ASTM D2166 or ASTM D2850).

3.2 Site Class Determination

Movement through the procedure can be generally broken into three steps as follows. A flowchart summarizing the steps is given in **Fig. G3–1**.

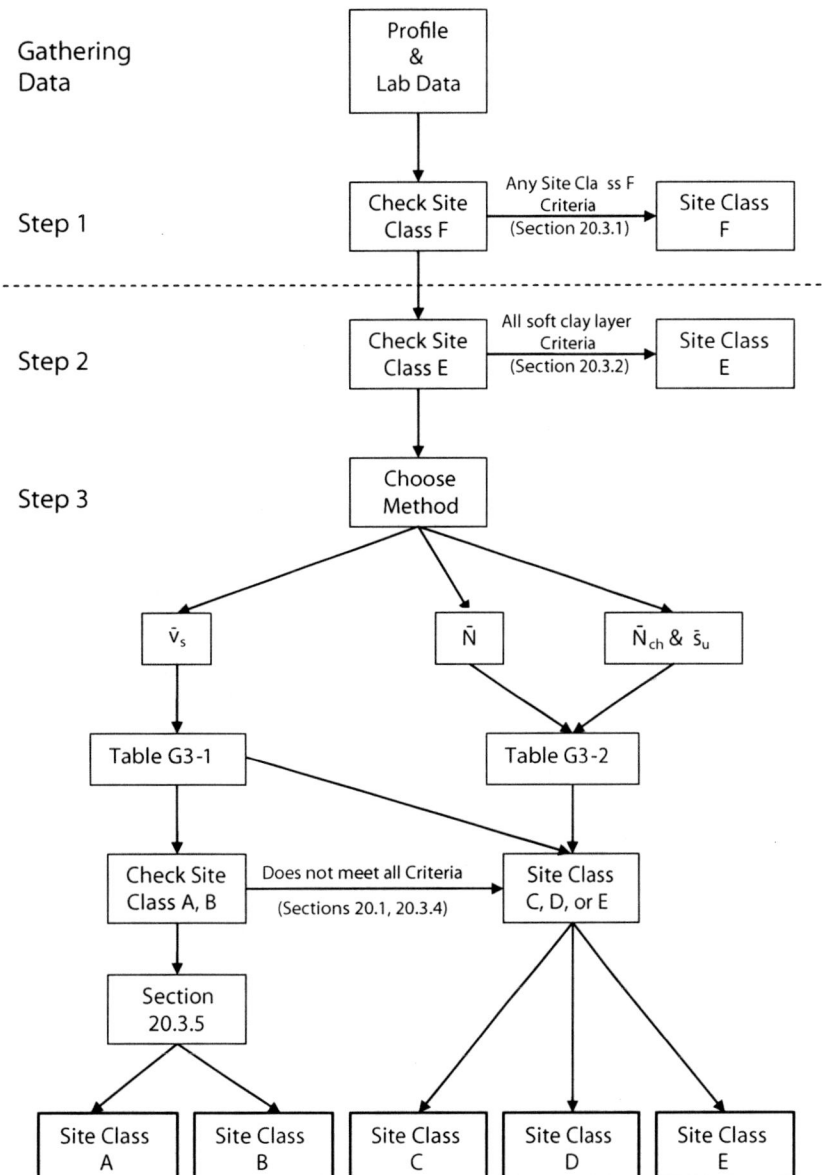

Figure G3–1 Site Class Determination Flowchart.

3.2.1 Step 1: Check Site Class F (Section 20.3.1)

If *any* of the following conditions are met, the site shall be classified as Site Class F:

1. Soils vulnerable to potential failure or collapse under seismic loading, such as

   - liquefiable soils,
   - quick or highly sensitive clays, or
   - collapsible or weakly cemented soils.

Exception: Fundamental building period $T \leq 0.5$ s (Section 20.3.1)
or
2. Peat and/or highly organic clays ($H > 10$ ft (3 m))
or
3. Very high plasticity clays ($H > 25$ ft (7.6 m) with PI > 75)
or
4. Very thick soft to medium stiff clays ($H > 120$ ft (37 m)) with $s_u < 1,000$ lb/ft^2 (50 kPa).

A site response analysis (Section 21.1) shall be performed for sites determined to be Site Class F.

3.2.2 Step 2: Check Site Class E (Section 20.3.2)

If a profile contains a soft clay layer with *all* of the following characteristics, the site shall be classified as Site Class E:

1. $H_{layer} > 10$ ft (3 m),
and
2. PI > 20,
and
3. $w \geq 40\%$,
and
4. $s_u < 500$ lb/ft^2 (25 kPa).

3.2.3 Step 3: Check Site Class C, D, E (Section 20.3.3), or A, B (Sections 20.3.4 and 20.3.5)

Using one of the following three methods, categorize the site using **Table G3–1**. All computations of $\overline{v}_s$, $\overline{N}$, $\overline{N}_{ch}$, $\overline{s}_u$ and shall be performed in accordance with Section 20.4.

$\overline{v}_s$ Method

An advantage of using shear wave velocity data is that the measured behavior better characterizes the subsurface profile than data collected from point locations (i.e., borings). Disadvantages of the method include the relatively expensive cost as well as experience required to perform and interpret the data.

If appropriate shear wave velocity data are available, $\overline{v}_s$ shall be calculated for the top 100 ft (30 m) using Eq. 20.4-1 and the appropriate Site Class determined from **Table G3–1**.

If the classification falls into criteria of either Site Class A or B in **Table G3–1**, the following additional criteria shall be considered:

- Site Class A or B shall not be assigned to a site if there is more than 10 ft (3 m) of soil between the rock surface and the bottom of the spread footing or mat foundation (Section 20.1).
- Shear wave velocity criteria specified in Section 20.3. 4 for Site Class B , and in Section 20.3.5 for Site Class A, shall be observed.

The applicable Site Class depends on which of the above criteria are met.

$\overline{N}$ Method

Using standard field penetration values for all soil and rock layers, $\overline{N}$ shall be calculated for the top 100 ft (30 m) using Eq. 20.4-2. The following should be considered regarding standard field penetration values (Section 20.4.2):

- ASCE 7-05 states that standard penetration values as "directly measured in the field without corrections" should be used. The author believes that energy corrections based on the type of hammer used should be applied because this difference is fundamental in the values measured. For instance, the standard penetration values from an automatic hammer should be appropriately adjusted to safety hammer (N_{60}) values to account for the high efficiency of the automatic hammer.
- Eq. 20.4-2 requires a single N value for each distinct layer in the profile. Average or conservatively choose values from multiple borings to characterize each distinct layer.
- Use a maximum of 100 blows/ft; see Section 20.4.2 for discussion.
- Where refusal is met for a rock layer, N shall be taken as 100 blows/ft.

$\overline{N}_{ch}$ and $\overline{s}_u$ Method

Divide the 100-ft (30-m) profile into cohesionless and cohesive layers in accordance with the definitions presented in Section 20.3.3.

Using standard field penetration values for the cohesionless layers, $\overline{N}_{ch}$ shall be calculated using Eq. 20.4-3. The comments on standard field penetration values listed in the section entitled $\overline{N}$ Method apply to $\overline{N}_{ch}$.

Using undrained shear strength values for the cohesive layers, $\overline{s}_u$ shall be calculated using Eq. 20.4-4. As stated in Section 20.4.3, undrained shear strength values shall be determined in accordance with ASTM D2166 or ASTM D2850.

Determining a Site Class requires two steps, first using $\overline{N}_{ch}$ as the classification criterion, and second using $\overline{s}_u$ as the classification criterion, using **Table G3–1**. If the Site Classes differ, the site shall be assigned a Site Class corresponding to the softer soil (Section 20.3.3).

3.3 Site Classification Example

The site profile presented in this example represents highly idealized subsurface conditions. Interpretation of actual subsurface data and soil properties requires substantial judgment by the geotechnical professional. The site profile used in the example is shown in **Fig. G3–2**.

The field standard penetration values in the example represent N_{60} values obtained from a safety hammer. As noted in the $\overline{N}$ Method section, the author believes these are the appropriate values to be used in seismic site class determination. The steps previously outlined are applied to the given example below.

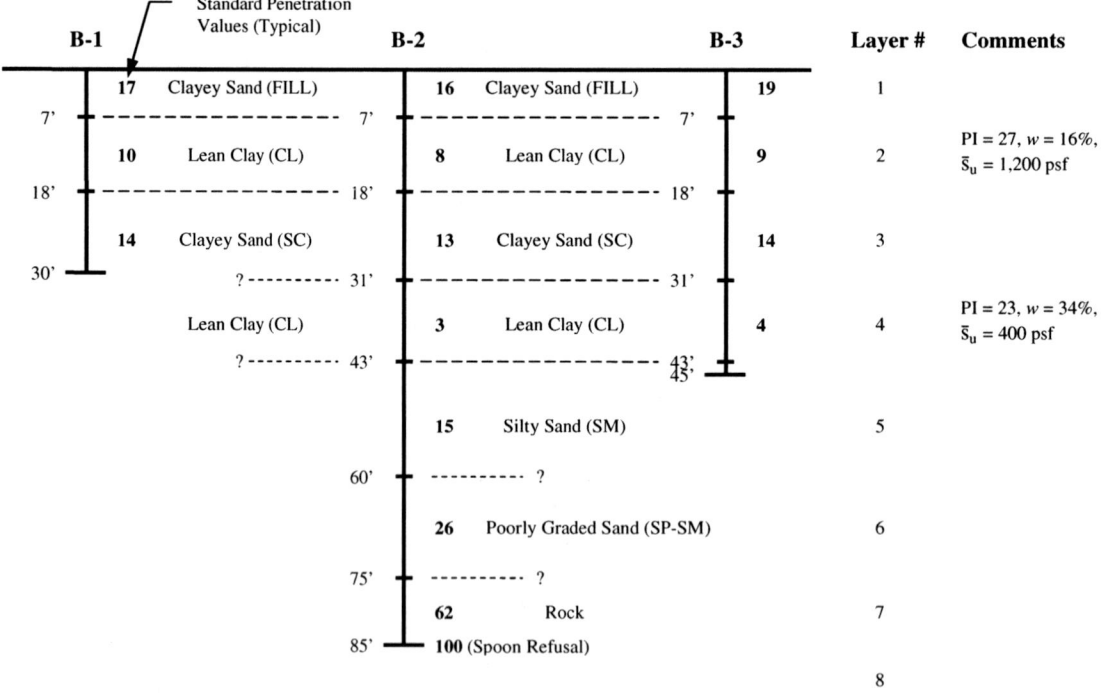

Figure G3–2 Subsurface Profile.
Notes: Standard Penetration Values presented as N_{60} values (ASTM 1586). Soil designations are based on USCS classification (ASTM 2487).

3.3.1 Step 1: Check Site Class F

If the profile meets *any* of the criteria in Section 20.3.1, the site shall be classified as Site Class F. This profile was selected to ensure that Site Class F does not apply. However, this check should not be overlooked in practice.

3.3.2 Step 2: Check Site Class E

If the profile contains any layers meeting *all* of the criteria in Section 20.3.2, the site shall be classified as Site Class E. Soft clay layer criteria are checked below:

Soft clay:	$H > 10$ ft	$PI > 20$	$w \geq 40\%$	$\overline{s}_u < 500$ lb/ft²
Layer 2:	$H = 11$ ft	$PI = 27$	**$w = 16\%$**	**$\overline{s}_u = 1{,}200$ lb/ft²**
Layer 4:	$H = 12$ ft	$PI = 23$	**$w = 34\%$**	$\overline{s}_u = 400$ lb/ft²

where bold indicates a criterion that is not met.

Layer 2 does not qualify based on its water content and undrained shear strength. Layer 4 does not qualify based on its water content.

Because neither layer satisfies *all* of the soft clay layer criteria, the site does not automatically qualify for Site Class E.

3.3.3 Step 3: Check Site Class C, D, E, or A and B

In this example, we use the $\overline{N}$ method to determine Site Class. Using this method automatically excludes Site Class A or B because they are based on shear wave velocity.

Using the notation from Eq. 20.4-2, the information in **Table G3–2** was determined from the example profile in **Fig. G3–2**. Based on the value of $\overline{N}$ = 100/8.65 = 12 calculated using Eq. 20.4-2, the site classifies as Site Class E ($\overline{N} < 15$) in accordance with **Table G3–1**.

Some observations that can be made from the values in **Table G3–2** include:

- Standard penetration value of 50 blows/in. at refusal (85 ft) was assigned a maximum value allowed of 100 blows/ft (Section 20.4.2).
- Based on the known geology, this blow count was then applied from refusal to a depth of 100 ft to complete the site profile, resulting in a 15-ft layer with blow counts of 100 blows/ft.

3.4 Comments on Site Classification

Although highly idealized, the site profile used in this example illustrates the need for adequate site investigation. Had only the data from shallow boring (B-1) been available, the designer would be unaware of the potentially soft clay layer encountered in B-2 and B-3.

Table G3–2 Summary of $\overline{N}$ Method

Layer No. i	Soil or Rock Designation	Cohesionless[a]	Cohesive[a]	N_i (blows/ft)	d_i (ft)	d_i/N_i
1	SC (FILL)	X		17	7	0.41
2	CL		X	9	11	1.22
3	SC	X		13	13	1.00
4	CL		X	3	12	4.00
5	SM	X		15	17	1.13
6	SP-SM	X		26	15	0.58
7	Rock	X		62	10	0.16
8	Rock	X		100	15	0.15
					Total 100	Total 8.65

$$\overline{N} = 100/8.65 = 12$$

a. Based on ASCE 7-05 definition; Section 20.3.3

Other important considerations that are not explicitly covered by the code include

- where to begin the site profile for below-grade structures,
- how to incorporate planned site grading (cut and fill),
- applicability of the design response spectrum method (generalized simplified seismic analysis) to structures supported on deep foundations, and
- characterizing highly variable site profiles (e.g., layer thickness and/or properties) within a given site.

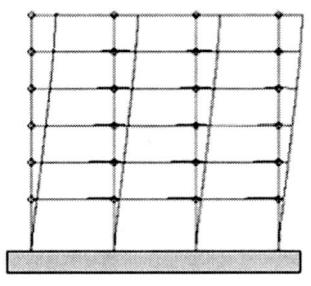

Example 4
Determining Ground Motion Parameters

In this example, the design basis spectral accelerations S_{DS} and S_{D1} are found for a site in Savannah, Georgia. They are first determined by hand, using maps and tables provided by ASCE 7, and are then checked through use of a software utility provided by the U.S. Geological Survey.

The basic ground motion parameters in ASCE 7 are S_S and S_1. S_S is the "short period" spectral acceleration ($T = 0.2$ s), and S_1 is the one-second ($T = 1.0$ s) spectral acceleration for sites on firm rock (Site Class B). These accelerations are based on the Maximum Considered Earthquake (MCE), for which there is a 2 percent probability of being exceeded in 50 years.

S_S and S_1 are used in a number of ways in ASCE 7, the most important of which is in the determination of the design-level acceleration parameters S_{DS} and S_{D1}. The design accelerations include a site coefficient factor (F_a or F_v), which accounts for soil characteristics different from firm rock, and an additional factor of 2/3, which effectively converts from the MCE basis to a somewhat lower level of shaking, called the Design Basis Earthquake (DBE).

S_S and S_1 are obtained from maps (Figs. 22-1 through 22-14) that are provided in Chapter 22. Alternately, the mapped accelerations and the site factor coefficients may be obtained from a computer program provided by the U.S. Geological Survey (USGS). This example illustrates the use of the maps and tables, and is then reworked using the USGS program.

4.1 Example: Find Ground Motion Values for a Site in Savannah, Georgia

For this example, we consider a site on a Site Class D soil in downtown Savannah, Georgia. ASCE 7 provides detailed maps for this part of the country in Fig. 22-9. Savannah lies on the coast, just below the border between South Carolina and Georgia. The border is shown by the broken line on the detailed maps. Savannah is also close to the intersection of the 32-deg latitude and 81-deg longitude lines that are shown in Fig. 22-9. The location is shown by the small circle in the bottom part of **Fig. G4–1**, which is the 0.2-s spectral response acceleration map from Fig. 22-9.

From the upper contour map in Fig. 22-9, it appears that Savannah lies close to the 0.4-g contour, so an acceleration of 0.4 g is used for S_S. From the lower map in Fig. 22-9, it appears that Savannah is somewhat closer to the 0.1-g contour than it is to the 0.15-g contour, so a value of 0.12 g will be used for S_1. Summarizing, we use the following values for Savannah:

$S_S = 0.40\ g$
$S_1 = 0.12\ g$

The site coefficients F_a and F_v are taken from Tables 11.4-1 and 11.4-2, respectively. Using Table 11.4-1, we see that for Site Class D it will be necessary to interpolate between values of $F_a = 1.6$ for $S_S = 0.25\ g$, and $F_a = 1.4$ for $S_S = 0.50\ g$. Referring to the left part of **Fig. G4–2**, we see that $F_a = 1.48$. Interpolation is also required to determine F_v. Using the right part of **Fig. G4–2**, it is seen that $F_v = 2.32$. Using these site coefficients, the site-amplified ground motion parameters are computed as follows:

$$S_{MS} = F_a S_S = 1.48(0.40) = 0.592\ g \qquad (11.4\text{-}1)$$
$$S_{M1} = F_v S_1 = 2.32(0.12) = 0.278\ g \qquad (11.4\text{-}2)$$

Although interpolating from Tables 11.4-1 and 11.4-2 is not difficult, it is somewhat inconvenient. For this reason, a variant of the tables is provided in **Appendix A** of this book in **Tables GA-1** and **GA-2**, in which the coefficients in the original tables are replaced by interpolation formulas. The example is reworked as follows:

from **Fig. GA–1** and **Table GA-1**: $F_a = 1.8 - 0.8\ S_S = 1.8 - 0.8(0.4) = 1.48$
from **Fig. GA–2** and **Table GA-2**: $F_v = 2.8 - 4.0\ S_1 = 2.8 - 4.0(0.12) = 2.32$

These are the same values determined from interpolation.

Given the site amplified ground motion values, the design ground motion values are obtained as follows:

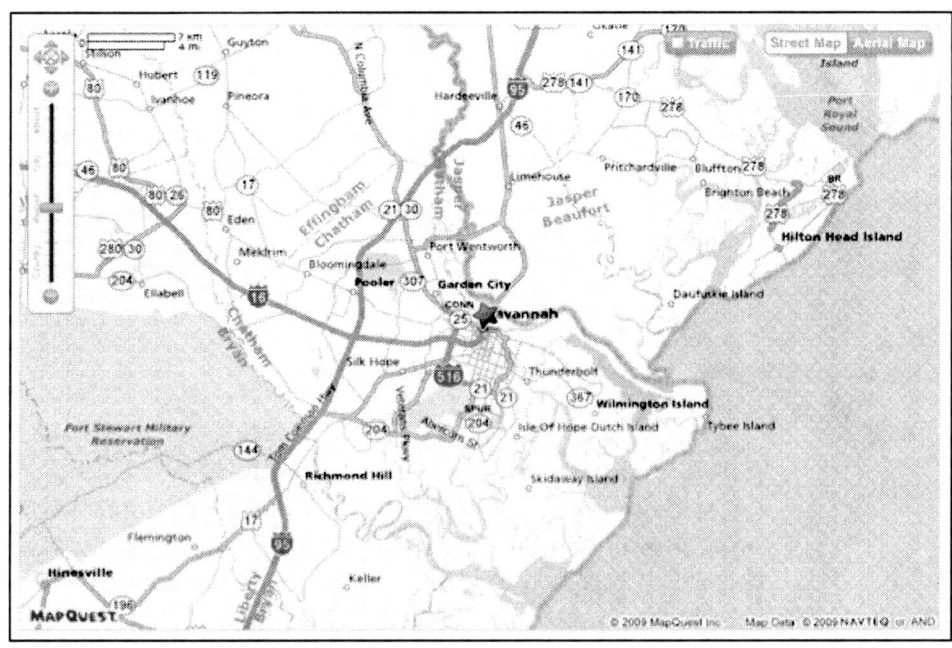

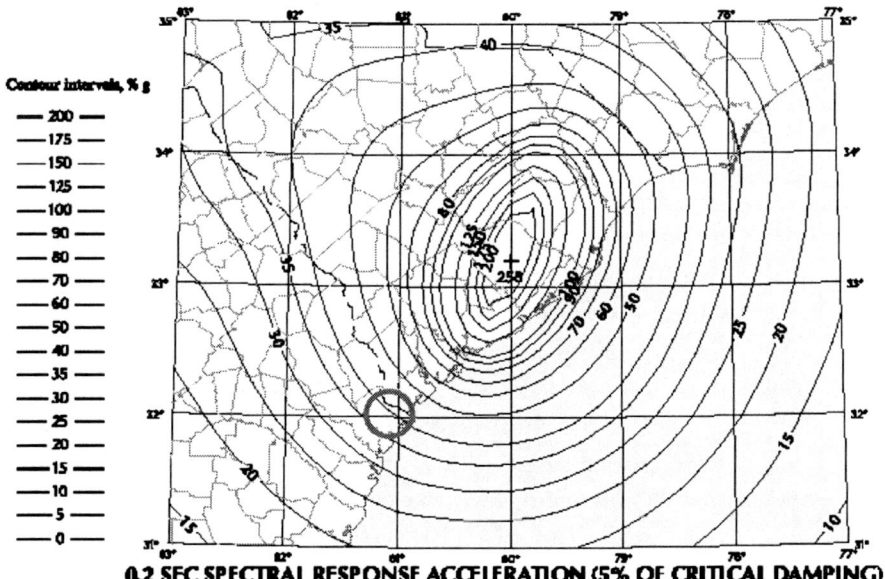

Figure G4–1   Savannah, Georgia, and Its Location on the ASCE 7 Map for S_s.

$$S_{DS} = (2/3) S_{MS} = (2/3)(0.592) = 0.395\ g \qquad (11.4\text{-}3)$$
$$S_{D1} = (2/3) S_{M1} = (2/3)(0.278) = 0.185\ g \qquad (11.4\text{-}4)$$

4.2 Use of the USGS Ground Motion Calculator

Even though detailed contour maps are provided for several regions of the country, it is usually more convenient to obtain the spectral ordinates S_S and

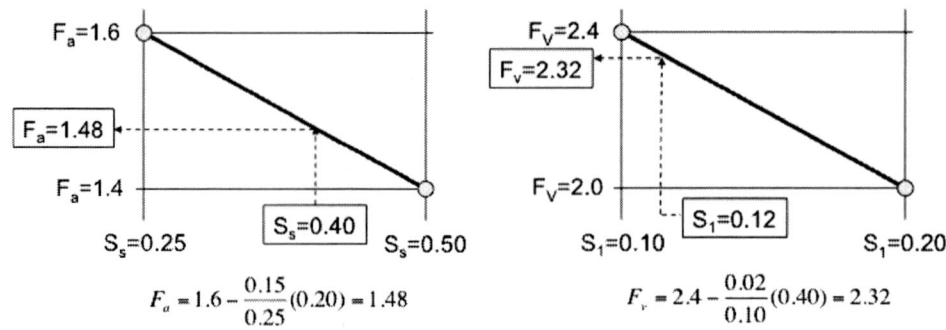

Figure G4–2 Interpolating for Site Coefficients F_a and F_v.

S_1 from a website and the Ground Motion Calculator (a Java application) that is maintained by the USGS. The web page containing the calculator has the following address: http://earthquake.usgs.gov/research/hazmaps/design/. If this link does not work, the site can usually be found by searching with the parameter "USGS Seismic Hazard Maps." **Appendix B** of this guide has detailed information regarding the USGS utility.

The main screen for the Ground Motion Calculator is shown in **Fig. G4–3**. At the top of the screen, the analysis option "ASCE 7 Standard..." has been selected. The 2005 ASCE 7 Standard is also selected in the box titled "Data Edition." Options are also available for use with other reference documents.

To obtain mapped values of S_S and S_1, either the latitude and longitude or the ZIP code of the site may be entered. The database does not contain all ZIP codes, so the use of latitude and longitude is often more convenient. Additionally, the use of latitude and longitude is more accurate for areas in which the area covered by a given ZIP code is large. Several Web-based utilities may allow the latitude and longitude to be found by specifying the physical address of the site. One such site is http://geocoder.us/.

If more than one set of spectral acceleration values is required at a time, you may provide a batch file with several latitude and longitude values. When using latitude and longitude, it is important to note that longitude values must be entered as a negative number (because the site is west of the Prime Meridian).

The USGS utility also has the capability to calculate the site class factors F_a and F_v and then the resulting design acceleration values S_{DS} and S_{D1}. Plots of the design response spectra (Fig. 11.4-1), in a variety of formats, are also available through the utility.

The use of the USGS Ground Motion Calculator is illustrated for the example considered above, which is in downtown Savannah, Georgia. The latitude and longitude ordinates for the Site Class D site, near the intersection of Anderson Street and East Broad Street are as follows:

latitude = 32.06 deg
longitude = –81.09 deg

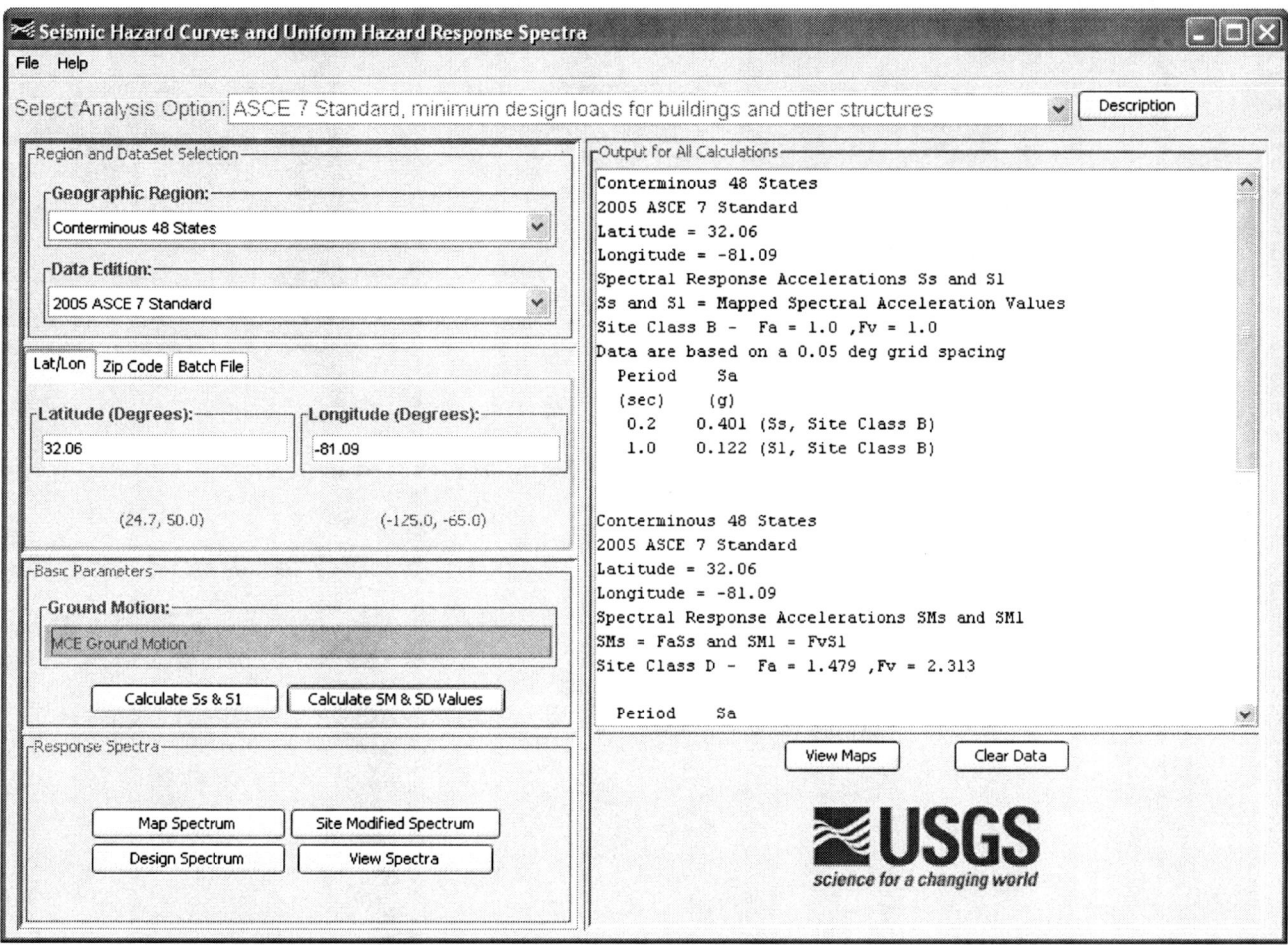

Figure G4–3 Ground Motion Parameters Computed for Site Class D Location in Savannah, Georgia.

The calculator (see **Fig. G4–3**) provides the following values:

$S_S = 0.401\ g$ $\qquad\qquad$ $S_1 = 0.122\ g$
$F_a = 1.479$ $\qquad\qquad$ $F_v = 2.313$
$S_{MS} = 0.593\ g$ $\qquad\qquad$ $S_{M1} = 0.282\ g$
$S_{DS} = 0.395\ g$ $\qquad\qquad$ $S_{D1} = 0.188\ g$

These values are very close to the values computed by hand.

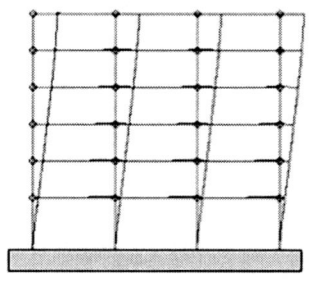

Example 5
Developing an Elastic Response Spectrum

In this example, an elastic response spectrum is generated for a site in Savannah, Georgia. An elastic spectrum is used directly in the Modal Response Spectrum analysis approach (Section 12.9), and as a basis for ground motion scaling in Response History analysis (Chapter 16).

For this example, we will use the same location in Savannah, Georgia, that was the setting for Example 4. This site is in downtown Savannah and is of Site Class D. The basic ground motion parameters, obtained through the use of the USGS Ground Motion Calculator, are summarized below:

$S_S = 0.401\ g$ $\qquad$ $S_1 = 0.122\ g$
$F_a = 1.479$ $\qquad$ $F_v = 2.313$
$S_{MS} = 0.593\ g$ $\qquad$ $S_{M1} = 0.282\ g$
$S_{DS} = 0.395\ g$ $\qquad$ $S_{D1} = 0.188\ g$

The basic form of the design response spectrum is shown in Fig. 11.4-1 of ASCE 7. This spectrum has four branches:

1. a straight-line ascending portion between $T = 0$ and $T = T_0$ (Eq. 11.4-5),
2. a constant acceleration portion between $T = T_0$ and $T = T_S$ ($S_a = S_{DS}$),
3. a descending "constant velocity" region between $T = T_S$ and $T = T_L$ (Eq. 11.4-6), and
4. a descending "constant displacement" region beyond T_L (Eq. 11.4-7).

The four branches of the spectrum are controlled by the design-level spectral accelerations S_{DS} and S_{D1}, by Eq. 11.4-5 (which describes the first branch of the spectrum), and by the "long-period transition period" T_L, which is provided by the contour maps of Figs. 22-15 through 22-20. For Savannah, the long-period transition, obtained from Fig. 22-15, is equal to 8 s. Such a long period would apply only for tall or flexible buildings or for the sloshing of fluid in tanks.

The other transitional periods, T_0 and T_S, are computed according to Section 11.4.5 as follows:

$T_S = S_{D1}/S_{DS} = 0.188/0.395 = 0.476$ s
$T_0 = 0.2 T_S = 0.095$ s

The spectral acceleration at $T = 0$ is given by Eq. 11.4-5. When $T = 0$, this equation produces an acceleration of $0.4(S_{D1}) = 0.4(0.395) = 0.158\ g$. This result is an approximation of the design-level peak ground acceleration.

The complete response spectrum for the Site Class D location in Savannah is plotted with a bold line in **Fig. G5–1**. The spectrum is plotted for a maximum period of 4.0 s, and as such, the fourth "constant displacement" branch is not shown.

For use in a computer program, the response spectrum is often presented in a table of period-acceleration values. In some cases, the spectrum is automatically generated from values of S_{DS} and S_{D1}.

When providing a table of spectrum values, it is important to provide sufficient resolution in the curved portions. Discrete spectral values are provided in **Table G5–1** for the Savannah site. These points are also represented by square symbols on the Site Class D spectrum on **Fig. G5–1**.

It is important to recognize that the elastic response spectrum developed in this example has not been adjusted by the importance factor I, nor by the response modification coefficient R. The use of these parameters, as well as the deflection amplification parameter C_d, are described in Section 12.9.2 of ASCE 7.

For illustrative purposes only, response spectra are also shown in **Fig. G5–1** for Site Classes A, B, C, and E. Parameters used to draw the curves are shown in **Table G5–2**. **Fig. G5–1** shows that site class can have a profound effect on the level of ground acceleration for which a structure must be designed. For the example given, S_{DS} for Site Class D is 1.48 times the value when the site class is B. S_{D1} increases by a factor of 2.32 when the site class changes from B to D.

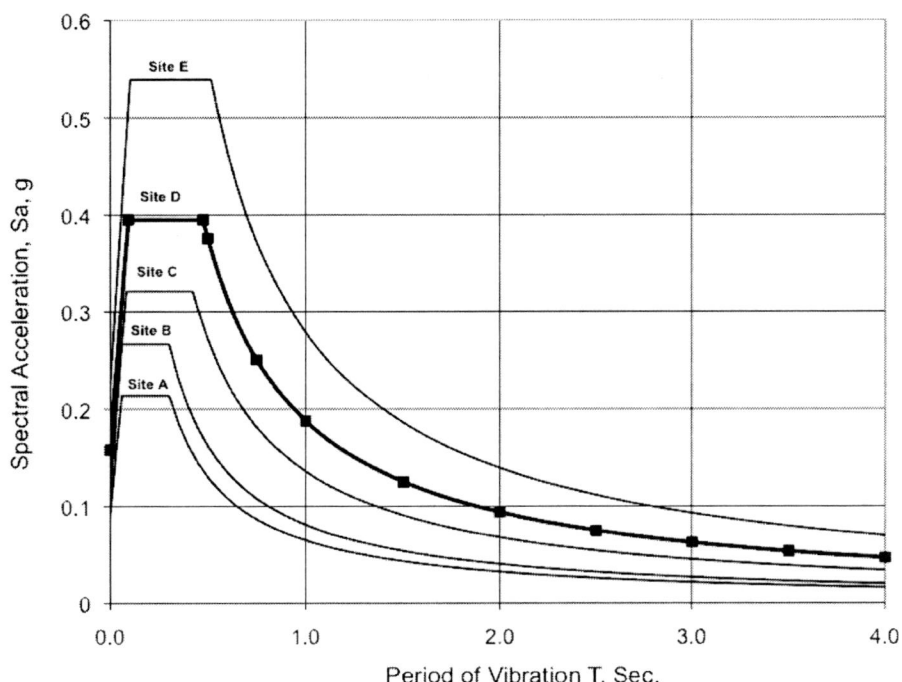

Figure G5-1 Elastic Design Response Spectra (R = 1, I = 1) for Various Site Classes Where $S_S = 0.401$ g and $S_1 = 0.122$ g

Table G5-1 Elastic Design Spectral Ordinates (R = 1, I = 1) for a Site Class D Location in Savannah, Georgia

Period T, s	Spectral Acceleration S_a (g)
0	0.158
0.095 (T_0)	0.395
0.476 (T_s)	0.395
0.50	0.376
0.75	0.251
1.00	0.188
1.50	0.125
2.00	0.094
2.50	0.075
3.00	0.063
3.50	0.054
4.00	0.047

Table G5–2 Elastic Design Response Spectrum Parameters (R = 1, I = 1) for Various Site Classes where $S_s = 0.401$ g, $S_1 = 0.122$ g

Site Class	F_a	F_v	S_{DS} (g)	S_{D1} (g)
A	0.8	0.8	0.214	0.065
B	1.0	1.0	0.267	0.081
C	1.20	1.678	0.321	0.136
D	1.479	2.312	0.395	0.188
E	2.017	3.434	0.539	0.279

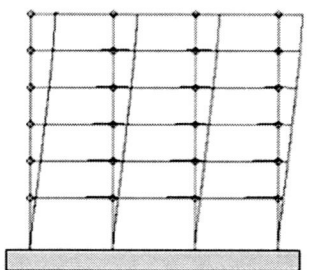

Example 6

Ground Motion Scaling for Response History Analysis

Chapter 16 of ASCE 7 provides the requirements for performing a linear or a nonlinear response history analysis. Among the key components of such an analysis are the selection of an appropriate suite of ground motions and the scaling of these motions. This example emphasizes the scaling procedure and provides only minimal background on the ground motion selection process. The scaling procedures are demonstrated for both two-dimensional and three-dimensional analysis. The example does not, however, proceed to use the selected and scaled motions in an actual analysis.

The scaling procedure is applied to a building in south central California. The building is in Seismic Design Category D and is situated on Site Class D soils. The site is not within 10 km of any known fault, so only far-field ground motions are considered.

The design-level spectral accelerations (see Section 11.4.4) are as follows:

$S_{DS} = 0.5\ g$
$S_{D1} = 0.2\ g$

The scaling procedure is demonstrated for buildings with a variety of periods of vibration. Both two-dimensional (2D) and three-dimensional (3D) analyses are considered.

6.1 Selection of Ground Motions

Sections 16.1.3.1 and 16.1.3.2 cover ground motion selection and scaling for 2D and 3D analysis, respectively. In both cases, it is required that the ground motions be selected from actual records and that they have magnitude, fault distance, and source mechanism consistent with those that control the Maximum Considered Earthquake.

There are a variety of sources of recorded ground motions. For this example, the Next Generation Attenuation (NGA) record set, provided by the Pacific Earthquake Engineering Research (PEER) Center was used. This record set is an updated version of the PEER Strong Motion Database. The NGA records are available from http://peer.berkeley.edu/nga. The Web site provides a search engine that allows the user to find ground motions by name (e.g. Northridge), magnitude range, and distance. Each ground motion record set consists of two horizontal records and one vertical acceleration record. The vertical record is generally not used for analysis. The PEER NGA database contained more than 3,500 record sets when this example was prepared. Appendix C of this guide provides additional information on the use of the PEER NGA Web site and database.

Section 16.1.4 of ASCE 7 requires that at least three record sets be used in any analysis. If fewer than seven sets are used, the response parameters used for design are the maximum values obtained among all of the analyses. If seven or more records are used, the design may be based on the average values obtained from the analysis. Because of this requirement, there is likely to be a substantial benefit to using seven or more ground motions. To maintain simplicity in this example, however, only three ground motion sets are used.

The ground motions that were selected for the analysis are listed in **Table G6–1**, and ground motion parameters are presented in **Table G6–2**. Each of these motions is considered to be a far-field motion because the epicentral distance is greater than 10 km. Each of the motions was recorded in Site Class D soil, and source mechanisms (fault type) are as shown in **Table G6–1**. The magnitudes of the earthquakes, in the range of 6.5 to 6.7, are somewhat lower than that which could be expected for the Maximum Considered Earthquake (MCE). There are few MCE level records available, however, so these records have to suffice.

Five percent damped pseudoacceleration response spectra for the two horizontal components of motions of each earthquake are shown in

Table G6–1 Record Sets Used for Analysis

Earthquake	PEER NGA ID	Year	Magnitude	ASCE 7 Site Class	Fault Type	Epicentral Distance (km)
San Fernando	68	1971	6.6	D	Thrust	39.5
Imperial Valley	169	1979	6.5	D	Strike-slip	33.7
Northridge	953	1994	6.7	D	Thrust	13.3

Table G6–2 Record Set Maxima

Earthquake	Component 1			Component 2		
	Bearing (deg)	PGA (g)	PGV (in./s)	Bearing (deg)	PGA (g)	PGV (in./s)
San Fernando	090	0.210	7.45	180	0.174	5.85
Imperial Valley	262	0.238	10.23	352	0.351	13.00
Northridge	009	0.416	23.2	279	0.516	24.71

Figs. G6–1(a) through **G6–1(c)**. The spectra were generated using the *NONLIN* program. (*NONLIN* may be obtained at no cost by FTP from ftp://filebox.vt.edu/users/fcharney/NONLIN/. *NONLIN* can read the PEER database files if they are saved to a text file with an "nga" file extension, for example, northridge.nga.) It is clear from these spectra that the Northridge ground motion record is dominant.

6.2 Scaling for Two-Dimensional Analysis

For 2D analysis, the "strongest" components from each ground motion pair, in terms of peak ground acceleration, are used. These components are as follows:

- San Fernando 090
- Imperial Valley 352
- Northridge 279

Other choices are possible for selecting the component to use, such as peak ground velocity, pseudoacceleration at the structure's fundamental period, spectral shape, or the analyst's judgment and experience.

The pseudoacceleration response spectra and the average of the spectra for the strongest components are shown in **Fig. G6–2(a)** Section 16.1.3.1 requires that the ground motions "be scaled such that the average value of the 5 percent damped response spectra for the suite of motions is not less than the design response spectrum for the site for periods ranging from $0.2T$ to $1.5T$."

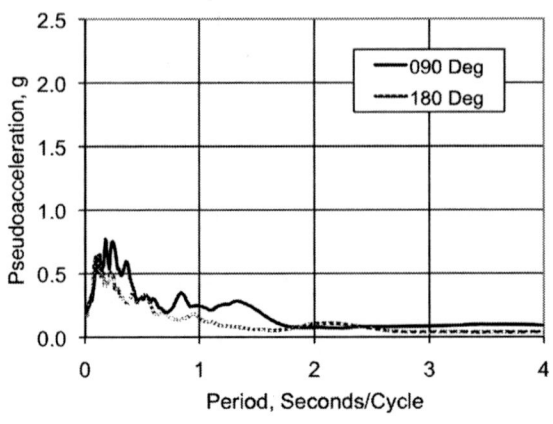

(a) San Fernando

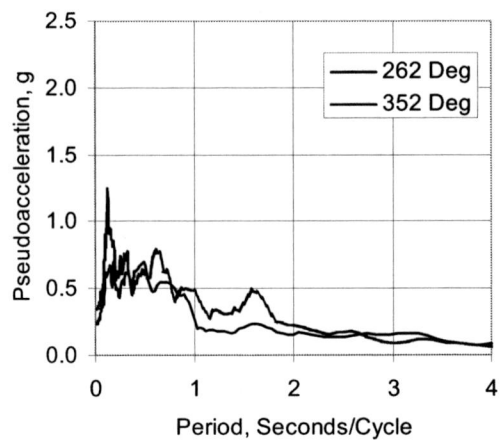

(b) Imperial Valley

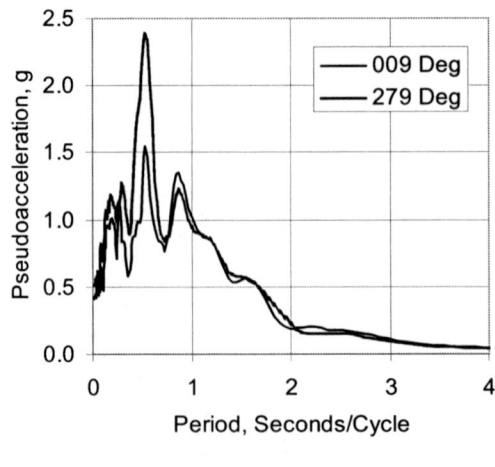

(c) Northridge

Figure G6–1 Pseudoacceleration Spectra for Selected Ground Motions.

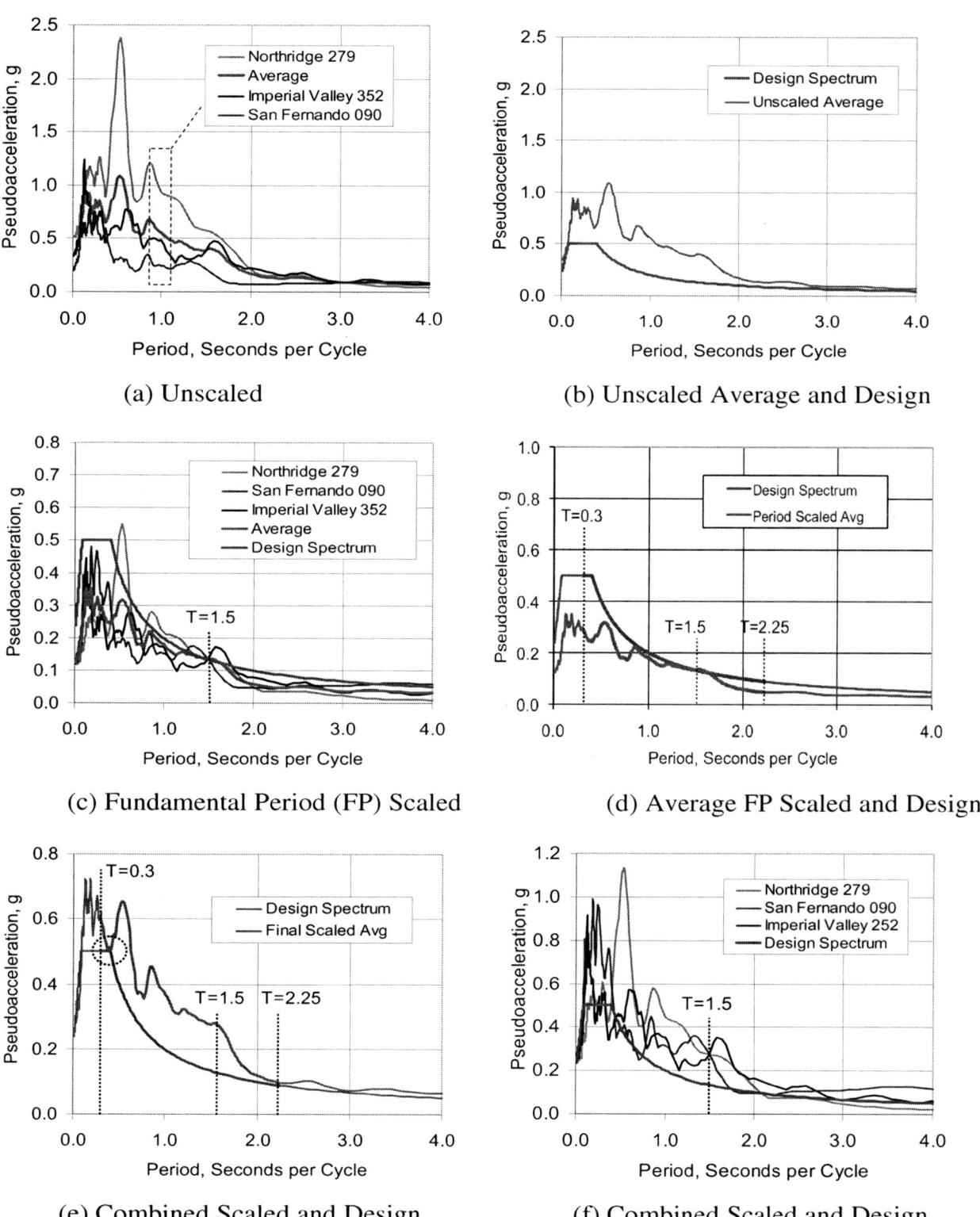

Figure G6-2 Various Spectra Resulting from the 2D Ground Motion Scaling Process.

The period would generally be determined from the same analysis model that will be eventually subjected to the response history analysis.

Given that each ground motion has its own scale factor, there are an infinite number of ways to scale the suite of motions such that the scaling requirements are met. In this example, a two-step scaling approach, which provides a unique set of scale factors, is used.

Step 1

Scale each ground motion such that it has the same spectral acceleration as the design spectrum at the structure's fundamental period of vibration. This step results in a different scale factor, FPS_i, for each motion i. For the purpose of this example, these motions are called the fundamental period scaled (FPS) motions.

Step 2

A second scale factor, called SS for suite scale factor, is applied to each of the FPS motions, such that the ASCE 7 scaling requirement is met. The combined scale factor, CS_i for each motion i, is SS times FPS_i. The procedure is applied to structures with a period, T, of 1.5 and 2.5 s.

Fig. G6–3(a) shows the unscaled spectra for each ground motion, as well as the average spectra. The number shown in the legend after the ground motion (e.g., 279 after Northridge) is the compass bearing of the ground motion. The dashed box at $T = 1.0$ s correlates the ground spectra shown in the plot with the ground motion identifier in the legend. For example, the Northridge earthquake is at the top of the legend and is the topmost curve in the dashed box. The unscaled average spectrum and the design spectrum (see Section 11.4.5 of ASCE 7) are shown in **Fig. G6–3(b)**. It is clear from this plot that the average ground motion spectrum falls above the design spectrum at all periods.

The spectral ordinates at $T = 1.5$ s are shown below for the design spectrum and for each of the individual ground motion spectra. The FPS factor for each ground motion is also shown. When these scale factors are applied to the ground motions, the resulting spectra have a common ordinate of 0.133 g at $T = 1.5$ s. This result is shown in **Fig. G6–3(c)**. The average of the FPS spectra are shown together with the design spectrum in **Fig. G6–3(d)**.

Design spectrum:	0.133 g	
San Fernando:	0.215 g	$FPS_1 = 0.133/0.215 = 0.619$
Imperial Valley:	0.373 g	$FPS_2 = 0.133/0.373 = 0.357$
Northridge:	0.578 g	$FPS_3 = 0.133/0.578 = 0.23$

The SS scale factor is determined such that no spectral ordinate in the combined average scaled ground motion spectrum falls below the design spectrum in the period range of 0.2 T to 1.5 T. With $T = 1.5$ s, this range is 0.3 to 2.25 s. The required SS scale factor is 2.06. As seen from **Fig. G6–3(e)**,

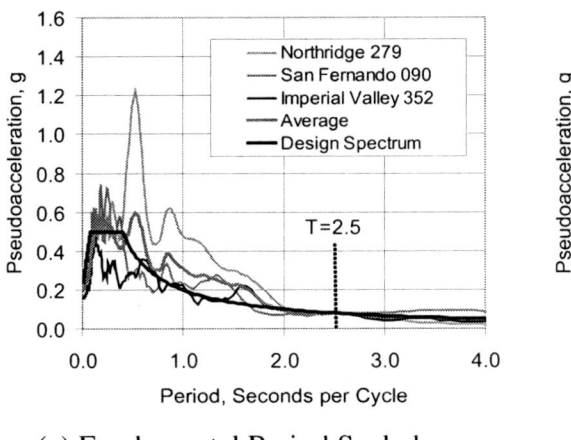
(a) Fundamental Period Scaled

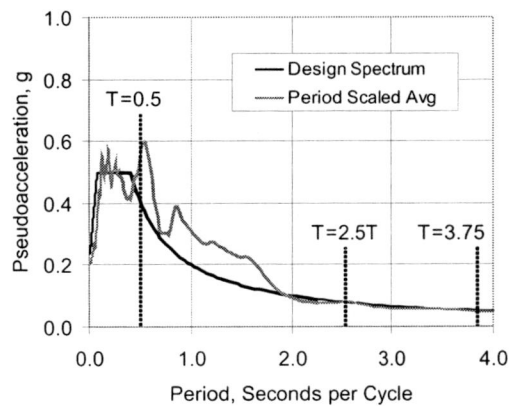
(b) Average FP Scaled and Design

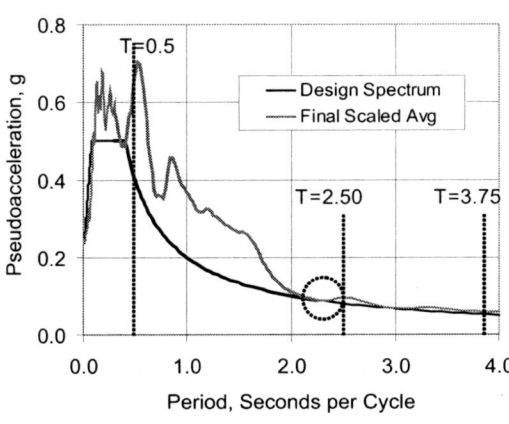
(c) Combined Scaled and Design

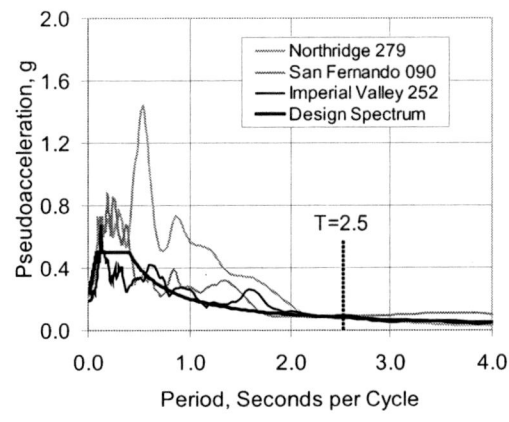
(d) Combined Scaled and Design

Figure G6–3 Various Spectra Resulting from the 2D Ground Motion Scaling Process.

the match point (inside the small dashed circle in **Fig. G6–3(e)** occurs at a period of about 0.4 s. The final combined scale factors for each ground motion are as follows:

San Fernando: $CS_1 = 0.619 \times 2.06 = 1.28$
Imperial Valley: $CS_2 = 0.357 \times 2.06 = 0.735$
Northridge: $CS_3 = 0.231 \times 2.06 = 0.476$

Fig. G6–3(f) shows the CS ground motions together with the design spectrum. The ground motion ordinates at the structure's fundamental period of vibration, 1.5 s, is 0.275 g for each earthquake, which is slightly more than two times the design spectrum ordinate (0.133 g) at the same period. It appears, therefore, that the scaling procedure has resulted in individual ground motions that are much stronger than that which would be implied by the design spectrum.

The scaling procedure is now repeated for a building with $T = 2.5$ s. The FPS factors are computed as follows:

Design spectrum: 0.0796 g
San Fernando: 0.0827 g $FPS_1 = 0.0796/0.0827 = 0.963$
Imperial Valley: 0.1729 g $FPS_2 = 0.0796/0.1729 = 0.460$
Northridge: 0.1550 g $FPS_3 = 0.0796/0.1550 = 0.513$

The resulting FPS spectra are shown in **Fig. G6–4(a)** where it may be seen that all spectra have a common ordinate of 0.0796 g at $T = 2.5$ s. The average FPS spectra and the design spectrum are shown in **Fig. G6–4(b)**. For a system with $T = 2.5$ s, the suite scale factor, SS, is 1.176, resulting in the following combined scale factors:

San Fernando: $CS_1 = 0.963 \times 1.176 = 1.13$
Imperial Valley: $CS_2 = 0.460 \times 1.176 = 0.541$
Northridge: $CS_3 = 0.513 \times 1.176 = 0.603$

The resulting FS average spectrum and the design spectrum are shown in **Fig. G6–4(c)**, where the match point, shown by the dashed circle, occurs at approximately $T = 2.25$ s. This result is in contrast to the scaling for the structure with $T = 1.5$ s (**Fig. G6–4(e)**), for which the match point was at approximately $T = 0.4$ s.

The individual CS spectra are shown in **Fig. G6–4(d)**. The ground motion spectra and the design spectrum have similar ordinates at the system's fundamental period of 2.5 s. However, the Northridge earthquake has significantly greater higher mode contributions as compared to the San Fernando or Imperial Valley ground motions.

It is clear from this example that the final combined scale factors are strongly dependent on the system's fundamental period and on the shape of the individual spectra. The dependence on individual spectra shape may be reduced somewhat by using a larger number of ground motions because the more ground motions used the smoother the average spectrum, and because the average spectral shape is less dependent on one dominant ground motion (e.g., the Northridge earthquake in this example). Additionally, the scaling approach demonstrated herein, where the first step in the scaling was the fundamental period scaling, removed a certain degree of freedom in scaling the individual ground motions. The advantage of the approach demonstrated is that any person scaling the same three ground motions obtains the same result. The disadvantage is loss of some control in the process. For example, there is nothing in ASCE 7 that says that the individual scale factors cannot be independently adjusted, as long as the final average ground motion spectrum does not fall below the design spectrum in the period range of 0.2 to 1.5 T.

6.3 Scaling for Three-Dimensional Analysis

The ground motion scaling requirements for 3D analysis are provided in Section 16.1.3.2. The scaling procedures are similar to those for 2D analysis, with the following exceptions:

1. For each earthquake in the suite, the square root of the sum of the squares (SRSS) of the spectra for each pair of horizontal components is computed. When computing the SRSS, the motion as recorded, without scale factors, is used.
2. Individual scale factors are applied to the SRSS spectra such that the average of the scaled SRSS spectra does not fall below 1.3 times the design spectra by more than 10 percent for any period between $0.2T$ and $1.5T$.

With regard to point 2, the period T is generally different in the two orthogonal directions, and thus, the scale factors would be different in the two directions.[1] Selection of the two periods for 3D analysis may not be straightforward for buildings in which there is a strong coupling of lateral and torsional response.

The scaling procedure is illustrated below for a building with a period of vibration of 1.5 s. As with the 2D approach, two scale factors are determined for each earthquake: a fundamental period scale factor FPS_i, which is unique for each earthquake, and a suite scale factor, SS, which is common to all earthquakes. The product of these two scale factors is the combined scale factor, CS_i.

The SRSS of the ground motion pairs and the average of the SRSS are shown in **Fig. G6–4(a)**. The average of the SRSS spectra is shown, together with 1.3 times the design spectrum in **Fig. G6–4(b)**. The first step in the scaling process is to scale each SRSS such that the spectral acceleration at the structure's fundamental period matches 1.3 times the design spectrum at the same period. The appropriate scale factors were determined as follows:

1.3 × design spectrum:	0.173 g	
San Fernando SRSS:	0.223 g	$FPS_1 = 0.173/0.223 = 0.776$
Imperial Valley SRSS:	0.428 g	$FPS_2 = 0.173/0.428 = 0.404$
Northridge SRSS:	0.797 g	$FPS_3 = 0.173/0.797 = 0.217$

The FPS spectra are shown with 1.3 times the design spectrum in **Fig. G6–4(c)**, and the average of the FPS spectra is shown with the design

---

1. Having different scale factors in the two different directions is not rational, and a different interpretation of the ASCE 7 requirements is warranted. One approach for handling different periods in different direction is to select the scaling range as $0.2T_{small}$ to $1.5T_{large}$, where T_{small} and T_{large} are the smaller and larger of the two first to fundamental periods of vibration. Another approach would be to select the scaling range as $0.2T_{avg}$ to $1.5T_{avg}$, where T_{avg} is the average of the two periods.

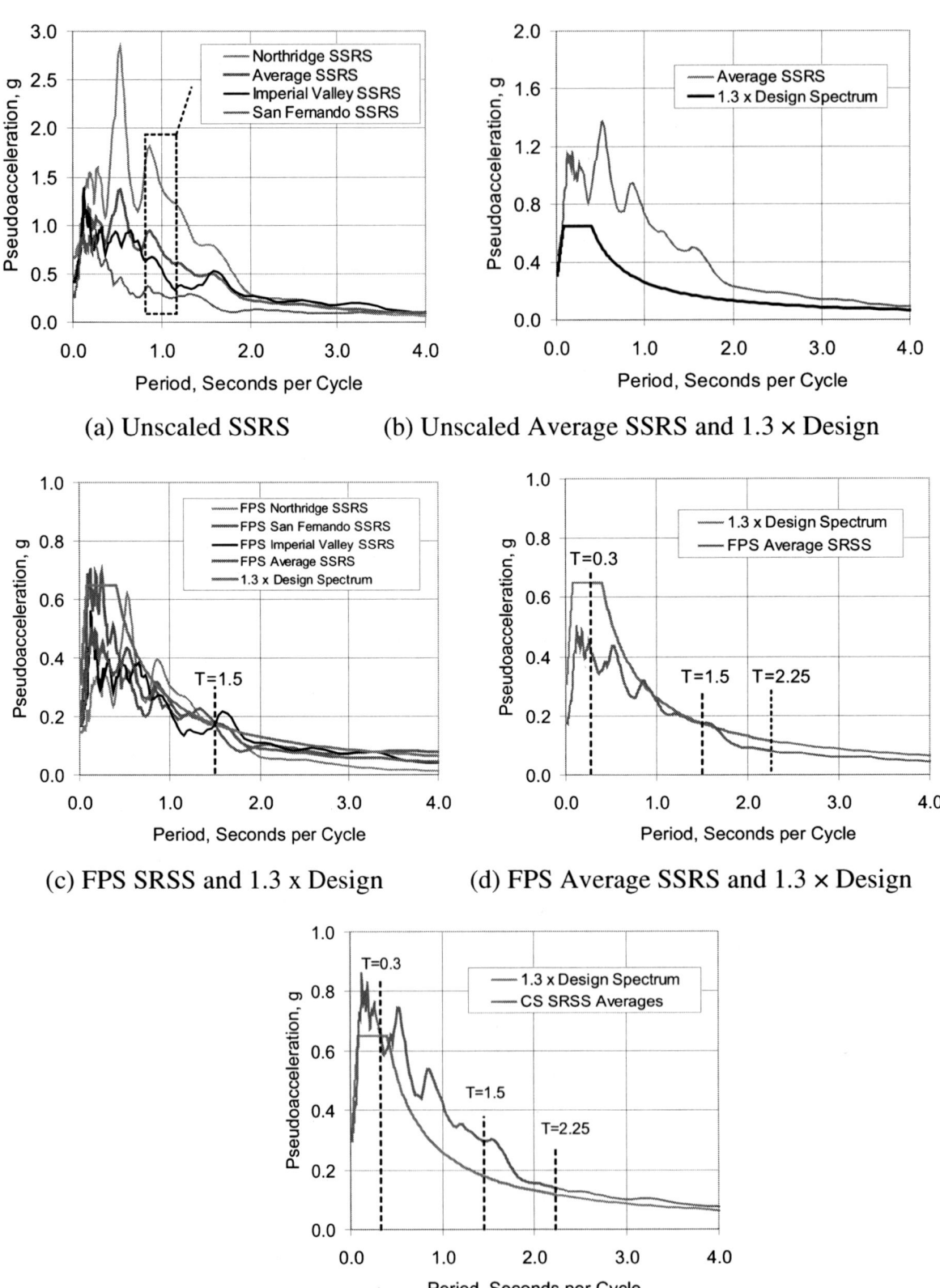

Figure G6–4 Various Spectra Resulting from the 3D Ground Motion Scaling Process.

spectrum in **Fig. G6–4(d)**. **Fig. G6–4(d)** also shows the period range over which the SS scale factor is to be determined.

The SS scale factor is computed such that no spectral ordinate in the combined average scaled SRSS ground motion spectrum falls below 1.3 times the design spectrum by more than 10 percent in the period range of $0.2T$ to $1.5T$. The required SS scale factor is 1.706. As seen from **Fig. G6–3(e)**, the controlling point occurs in the period range of 0.3 to 0.4 s. The final scale factors for each ground motion are as follows:

San Fernando SRSS: $CS_1 = 0.778 \times 1.706 = 1.38$
Imperial Valley SRSS: $CS_2 = 0.406 \times 1.706 = 0.693$
Northridge SRSS: $CS_3 = 0.219 \times 1.706 = 0.374$

The same scale factor (e.g., 1.38 for San Fernando) is applied to each component of the ground motion, and the two scaled components would be applied simultaneously in the analysis.

As with the 2D scaling, the 3D scaling results in highly amplified spectral ordinates in the range of the structure's 1.5-s fundamental period.

6.4 Comments on Ground Motion Scaling

This example has illustrated only one interpretation (the author's) of the ASCE 7 ground motion scaling requirements. Any analyst who uses the illustrated approach obtains the same set of scale factors for the same suite of ground motions. This result occurs because of the intermediate step in which the FPS factors are applied. ASCE 7 does not require this step, and for this reason, it is possible for different analysts to obtain different ground motion scale factors when the ASCE 7 procedure is used without the FPS scaling.

The 2D and 3D scaling for the structure with a period of 1.5 s produced final scale factors that have highly amplified spectral ordinates at the structure's fundamental period. This result occurs because of the fact that the SS scaling was controlled by the portion of the spectrum in the 0.3- to 0.4-s range. The amplification factors would likely have been somewhat lower if more than three ground motions were used in the analysis because the average ground motion spectrum becomes smoother as the number of ground motions increases.

It would, of course, also be possible to obtain more favorable scale factors by choosing different ground motions. With some effort, it might be possible to find three to seven ground motions for which the average spectrum is similar to the design spectrum in the period range of 0.2 to 1.5 times the fundamental period.

As a final note, ground motion selection and scaling are part of a complex process and should not be attempted without assistance from an experienced engineering seismologist.

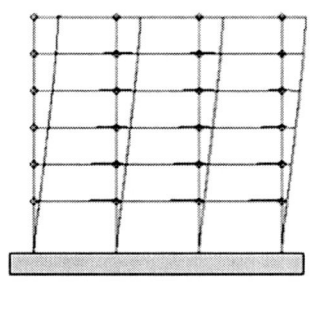

Example 7

Selection of Structural Systems

In this example, all viable structural steel systems for two buildings in an office complex are examined. The buildings have the same plan layout, but have different heights. One building is ten stories tall, and the other building is four stories tall. At the end of this example, a few comments are made in regard to bearing wall systems.

The best structural system for a particular building depends on a large number of factors, such as architectural and functional requirements, labor, fabrication, and construction costs. These issues are so variable that they cannot be addressed in this guide. Another consideration in system selection is how ASCE 7 restricts the kinds of systems that may be used. This example discusses these restrictions.

Section 12.2.1, together with Table 12.2-1, provides the basic rules for the selection of seismic force resisting systems. Table 12.2-1 is divided into broad categories, such as Bearing Wall Systems, Building Frame Systems, and Dual Systems, and it provides a number of system types within each category. Limitations are placed on the use of each system in terms of Seismic Design Category (SDC) and height. For example, ordinary steel concentrically braced frames (System 4 under the Building Frame Systems category) are not permitted (NP) in SDC F, are allowed only up to heights of 35 ft in SDC D and E, and are allowed with no height limit (NL) in SDC B and C. Note that Section 12.2.5.4 allows the tabulated height limits to be increased in certain circumstances.

For each system, three design parameters are specified:

- response modification coefficient R
- system overstrength factor Ω_o
- deflection amplification factor C_d

Whereas each of these parameters affects system economy, the most influential factor is R because the design base shear is inversely proportional to this parameter via the seismic response coefficient C_s. This coefficient is specified in Section 12.8. There are basically two sets of C_s equations. The first set of equations, given by Eqs. 12.8-2, 12.8-3, and 12.8-4, represent three branches of the inelastic design response spectrum. Each of these equations contains R in the denominator, so it appears that larger values of R result in lower values of design base shear.

The second set of equations provides the minimum values for C_s. The first of these, Eq. 12.8-5, is applicable when the mapped spectral acceleration, S_1, is less than 0.6 g. This equation

$$C_s = 0.044 S_{DS} I \geq 0.01$$

is not a direct function of R but will control over Eq. 12.8-3 whenever the fundamental period of vibration is greater than 22.7 T_S/R.

To show the effect of the base shear equations, the following example uses the building shown in **Fig. G7–1**. A four-story and a ten-story version of the building are considered. The four-story building has the same floor plan and first-story height as the taller building. Based on the building use, the Occupancy Category of the building is II (see Table 1-1). The structure is located in an area of relatively high seismicity and is situated on Site Class C soils. The ground motion parameters are as follows:

$S_S = 0.75\ g$	
$S_1 = 0.22\ g$	
$T_L = 6.0\ s$	
$F_a = 1.1$	(**Table GA-1** or Table 11.4-1)
$F_v = 1.58$	(**Table GA-2** or Table 11.4-2)
$S_{DS} = (2/3)(1.1)(0.75) = 0.55\ g$	(Eqs. 11.4-1 and 11.4-3)
$S_{D1} = (2/3)(1.58)(0.22) = 0.23\ g$	(Eqs. 11.4-2 and 11.4-4)
W (ten-story building) = 22,000 kips	
W (four-story building) = 8,800 kips	

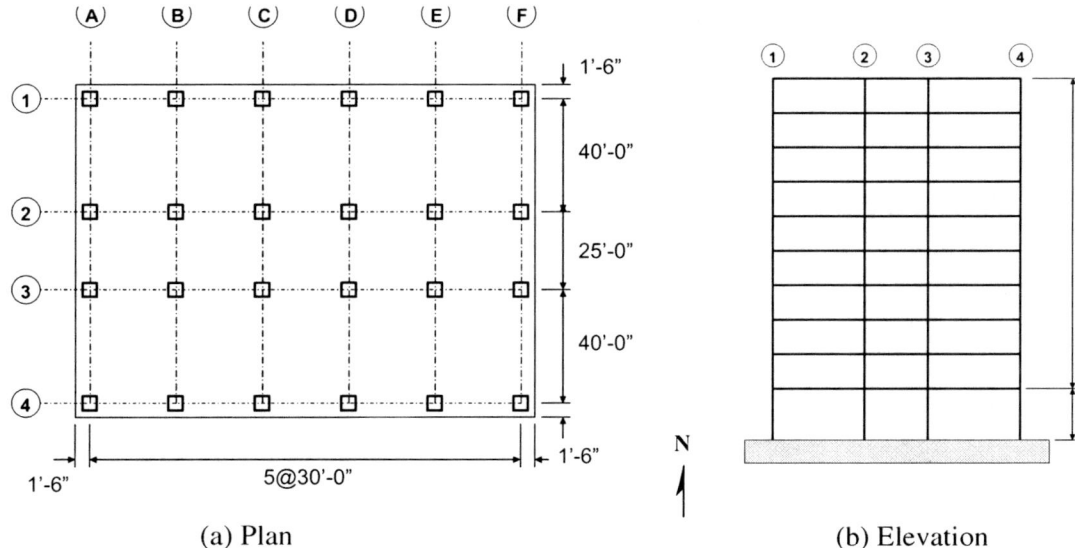

(a) Plan (b) Elevation

Figure G7–1 Base Building for System Comparison Examples.

Based on Table 11.5-1, the importance factor I is 1.0. In accordance with Section 11.6 and Tables 11.6-1 and 11.6-2, the building is assigned to SDC D.

A survey of locally available materials and labor costs has indicated that the structural system should be constructed from structural steel. Any structural system may be used, but architectural considerations require that no shear walls or diagonal bracing elements be used at the perimeter of the building.

Table G7–1 and **Table G7–2** list all of the viable structural systems for the ten-story and the four-story buildings, respectively. These systems are taken directly from Table 12.2-1 of ASCE 7. **Tables G7–1** and **G7–2** also list the system R and C_d values, the period of vibration, the seismic response coefficient C_s, the design base shear, and the effective R value, which is defined later in this example. The C_s values are given as computed using Eqs. 12.8-3 and 12.8-5. The period of vibration is taken as $T = C_u T_a$, with $C_u = 1.47$ for all systems. T_a is based on Eq. 12.8-7 and Table 12.8-2.

The base shear was computed using Eq. 12.8-1:

$$V = C_s W$$

A number of equations are provided for computing C_s. Equation 12.8-2 controls only if the building period is less than $T_S = S_{D1}/S_{DS} = 0.23/0.55 = 0.418$ s, which is not the case for either of the buildings examined. Eq. 12.8-4 is not applicable because none of the building periods exceed T_L, which is 6.0 s. Eq. 12.8-6 is not applicable because S_1 is < 0.6 g. This elimination leaves only Eqs. 12.8-3 and 12.8-5:

$$C_S = \frac{S_{D1}}{T\left(\dfrac{R}{I}\right)} \tag{12-8-3}$$

Table G7–1 Comparison of Structural Systems (10-Story Building)

1	2	3	4	5	6	7	8	9
System Number	Structural System	Height Limit (ft)	R	$C_u T_a$ (s)	C_s (Eq. 12.8-3)	C_s (Eq. 12.8-5)	V (kips)	R_{eff}
B-1	EBF with moment resisting connections	160	8	1.73	0.0166	0.0242	532	5.5
B-2	EBF with nonmoment resisting connections	160	7	1.73	0.0190	0.0242	532	5.5
B-3	Special concentrically braced frame	160	6	1.15	0.0333	0.0242	732	6.0
B-25	Buckling restrained brace with nonmoment resisting beam-column connections	160	7	1.73	0.0190	0.0242	532	5.5
B-26	Buckling restrained brace with moment resisting beam-column connections	160	8	1.73	0.0166	0.0242	532	5.5
B-27	Special steel plate shearwall	160	7	1.15	0.0285	0.0242	627	7.0
C-1	Special steel moment frame	No limit	8	2.06	0.0140	0.0242	532	4.6
C-2	Special steel truss moment frame	160	7	2.06	0.0160	0.0242	532	4.6
D-1	Dual system with steel EBF	No limit	8	1.15	0.0250	0.0242	550	8.0
D-2	Dual system with steel CBF	No limit	7	1.15	0.0285	0.0242	627	7.0
D-12	Dual system with BRB	No limit	8	1.15	0.0250	0.0242	550	8.0
D-13	Dual system with steel plate shearwall	No limit	8	1.15	0.0250	0.0242	550	8.0

Notes: The system number is the same as designated in Table 12.2-1 of ASCE 7. The period computed period for the special steel truss moment frames is assumed to be the same as that for a standard moment frame. EBF, eccentrically braced frame; CBF, concentrically braced frame; and BRB, buckling restrained brace.

$$C_S = 0.044 S_{DS} I \geq 0.01 \qquad (12.8\text{-}5)$$

Eq. 12.8-5 controls when $T > T_{MF}$, where

$$T_{MF} = 22.7\ T_S/R$$

Physically, T_{MF} is the period at which Eqs. 12.8-3 and 12.8-5 give the same values for C_s.

For the ten-story building, C_s given by Eq. 12.8-5 controls for the eccentrically braced frames, the buckling restrained braced frames, and the moment frames. This result occurs because of the high R values and the relatively long period of vibration for these systems. Eq. 12.8-5 does not control for the dual systems because these systems, which have high R values, have relatively low periods of vibration.

The effective R value given in column 9 of **Table G7–1** is the value of R that produces the controlling base shear. For example, for the first system

Table G7-2 Comparison of Structural Systems (4-Story Building)

1	2	3	4	5	6	7	8	9
System Number	Structural System	Height Limit (ft)	R	$C_u T_a$ (s)	C_s (Eq. 12.8-3)	C_s (Eq. 12.8-5)	V (kips)	R_{eff}
B-1	EBF with moment resisting connections	160	8	0.89	0.0323	0.0242	284	8
B-2	EBF with nonmoment resisting connections	160	7	0.89	0.0369	0.0242	324	7
B-3	Special concentrically braced frame	160	6	0.59	0.0646	0.0242	568	6
B-25	Buckling restrained brace with non-moment resisting beam-column connections	160	7	0.89	0.0369	0.0242	324	7
B-26	Buckling restrained brace with moment resisting beam-column connections	160	8	0.89	0.0323	0.0242	284	8
B-27	Special steel plate shearwall	160	7	0.59	0.0553	0.0242	487	7
C-1	Special steel moment frame	No limit	8	1.02	0.0283	0.0242	249	8
C-2	Special steel truss moment frame	160	7	1.02	0.0323	0.0242	285	7
D-1	Dual system with steel EBF	No limit	8	0.59	0.0484	0.0242	426	8
D-2	Dual system with steel CBF	No limit	7	0.59	0.0553	0.0242	487	7
D-12	Dual system with BRB	No limit	8	0.59	0.0484	0.0242	426	8
D-13	Dual system with steel plate shearwall	No limit	8	0.59	0.0484	0.0242	426	8

Notes: The system number is the same as designated in Table 12.2-1 of ASCE 7. The period computed period for the special steel truss moment frames is assumed to be the same as that for a standard moment frame. EBF, eccentrically braced frame; CBF, concentrically braced frame; and BRB, buckling restrained brace.

listed, the controlling base shear is 532 kips. Using Eqs. 12.8-1 and 12.8-3, R_{eff} is computed as follows:

$$R_{eff} = \frac{S_{D1} I}{VT} W = \frac{0.23(1.0)}{532(1.73)} 22,000 = 5.5$$

For the systems with the base shear controlled by Eq. 12.8-5, the effective R value is less than or equal to the tabulated R value (from Table 12.2-1).

For the four-story building (see **Table G7–2**), all of the base shears are controlled by Eq. 12.8-3, so the effective R values are the same as the tabulated values.

The point to make about the analyses shown in **Table G7–1** and **Table G7–2** is that for systems with high tabulated R values and relatively high periods, it is likely that Eq. 12.8-5 controls the base shear, and the apparent benefit of the high R value is lost.

There are also implications related to the calculation of drift. This issue is explored in Example 19. Additional commentary related to the minimum base shear is provided in Example 18.

7.1 Steel Frame Systems Not Specifically Detailed for Seismic Resistance

Part H of Table 12.2-1 provides design values for steel systems not specifically detailed for seismic resistance. These systems may be designed using the *Specification for Structural Steel Buildings* (AISC 2005c), and do not rely on the requirements of the *Seismic Provisions for Structural Steel Buildings* (AISC 2005a). In some cases it may be found that R = 3 systems are more economical than systems with higher R values (e.g., ordinary steel moment frames with R = 3.5) that are allowed for use in the same Seismic Design Category.

7.2 Bearing Wall Systems

Section 11.2 defines bearing wall systems (under the definition for "Wall") as systems in which bearing walls support all or major parts of the vertical load. Presumably, a major portion would be more than 50 percent of the total vertical load. Bearing walls are defined as (1) a "metal or wood stud wall that supports more than 100 lb/linear ft of vertical load in addition to its own weight," or (2) a "concrete or masonry wall that supports more than 200 lb/linear ft of vertical load in addition to its own weight." Given this definition, the likelihood is that most structural walls would be classified as bearing walls.

Consider the system shown in **Fig. G7–2**. This system, one story high, has precast concrete walls around the perimeter. The roof framing consists of steel interior tube columns and steel beams and joists. There are also steel columns between the walls on the east and west faces of the building, and these columns support the steel beams. The walls on the north and south side of the building, designated by "B," are clearly bearing walls. The walls on the east and west faces would be classified as bearing walls if the loading delivered into these walls by the roof deck exceeds 200 lb/linear ft. For the purpose of this example, it is assumed that this tributary loading is less than 200 lb/linear ft, and that these walls would not be designated as bearing walls, hence the designation "N" in **Fig. G7–2**.

The tributary vertical gravity load carried by the columns is shown by the shaded region in the interior of the structure. This region represents approximately 65 percent of the total load, so the bearing walls carry only 35 percent of the total vertical load. Hence, by definition, this system is not a bearing wall system and would be classified as a building frame system.

In **Fig. G7-3**, the system is changed such that there is only one line of columns in the interior, and the center walls on the east and west side of the building support reactions from the steel girder. Hence, these walls become bearing walls. The shaded region, representing the tributary vertical load carried by the columns, is slightly more that 50 percent of the total area, so the system would still be classified as a building frame system.

If the steel columns were removed in their entirety and the joists spanned the full width of the building, as shown in **Fig. G7-4** the building would be classified as a bearing wall system for loads acting in the east–west direction. For loads acting in the north–south direction, the walls resisting the lateral load are not bearing walls, and the system could be classified as a nonbearing wall system for loads acting in this direction.

The main difference between bearing wall systems and building frame systems (which encompass the same lateral load resisting elements) is that the R values for bearing wall systems are generally lower than those for the corresponding building frame system. Additionally, there are no bearing wall systems in structural steel, except for light-framed systems with steel sheets or light strap bracing.

Additional discussion related to bearing wall systems may be found in Ghosh and Dowty (2007).

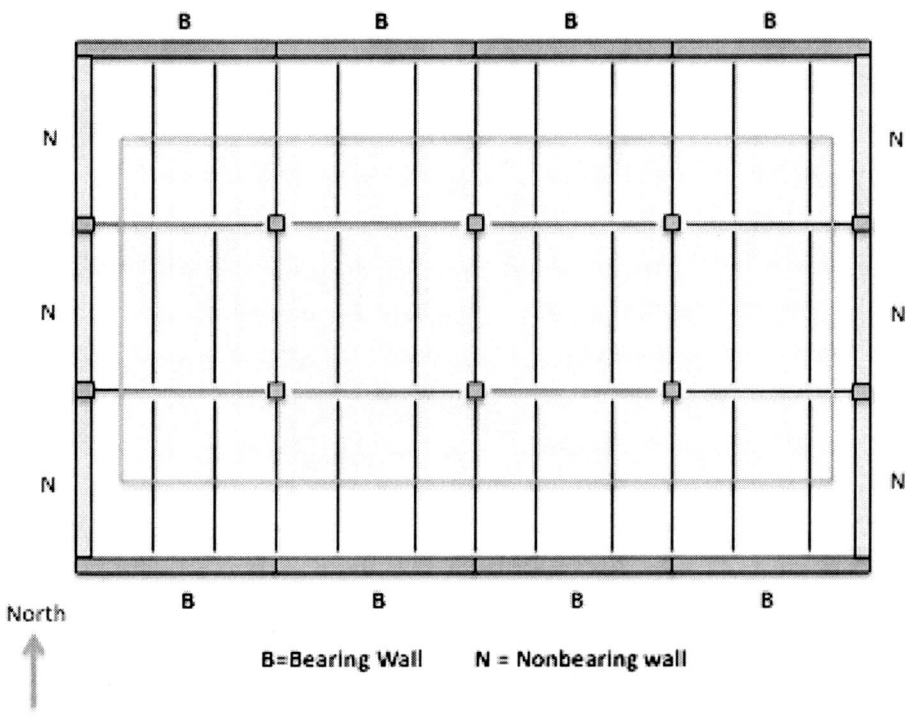

Figure G7-2 Concrete Shear Wall System, Scheme 1: Not a Bearing Wall System.

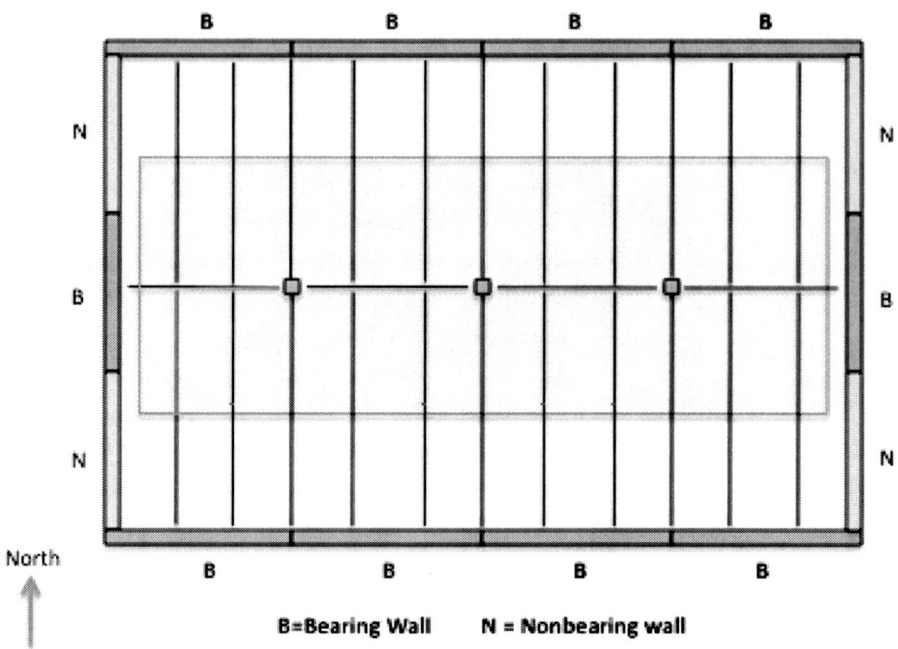

Figure G7–3 Concrete Shear Wall System, Scheme 2: Not a Bearing Wall System.

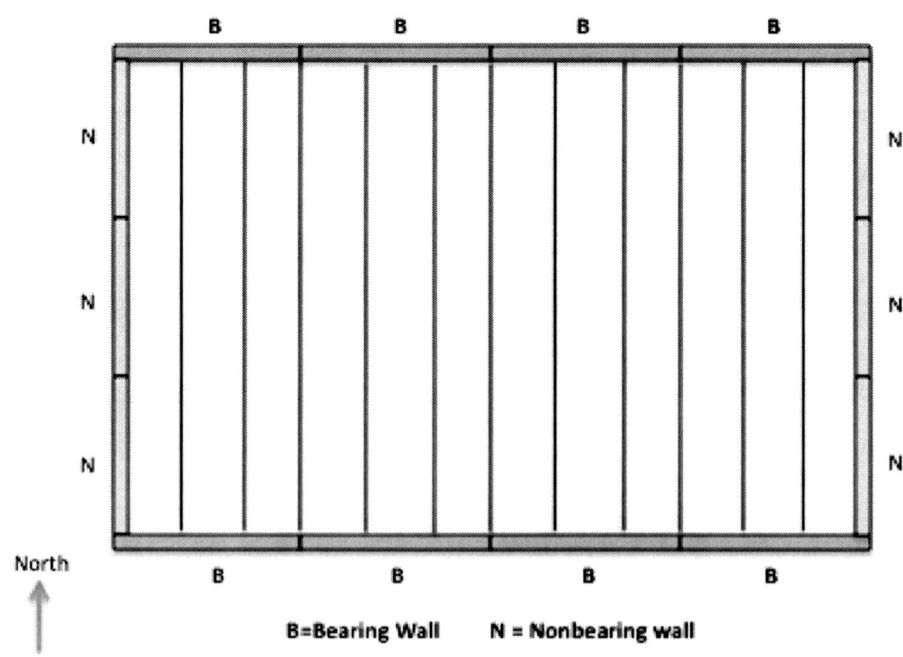

Figure G7–4 Concrete Shear Wall System, Scheme 3: Bearing Wall System.

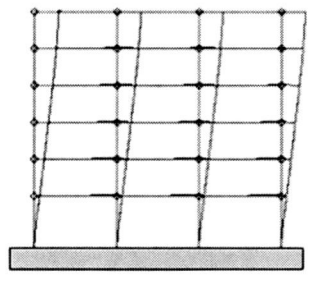

Example 8

Combinations of Lateral Load Resisting Systems

The examples in this chapter deal with the situation in which there are different lateral load resisting systems in the same direction, in orthogonal directions, or along the height of a building. At issue is the height limitation for the combined system and the appropriate values of R, Ω_o, and C_d to use for the systems.

8.1 Combinations of Framing Systems in Different Directions

Where different systems are used in the different (orthogonal) directions of a building, and where there is virtually no interaction between these systems, the system limitations set forth in Table 12.2-1 apply independently to the two orthogonal directions. The exception, of course, is that the lower of the two height limitations controls for the whole building.

Consider, for example, the buildings shown in **Fig. G8–1**, which are plan views of simple buildings with special reinforced concrete (RC) moment frames, special RC shear walls, or a combination of systems. The lateral load systems shown extend the full height of the buildings. Each of

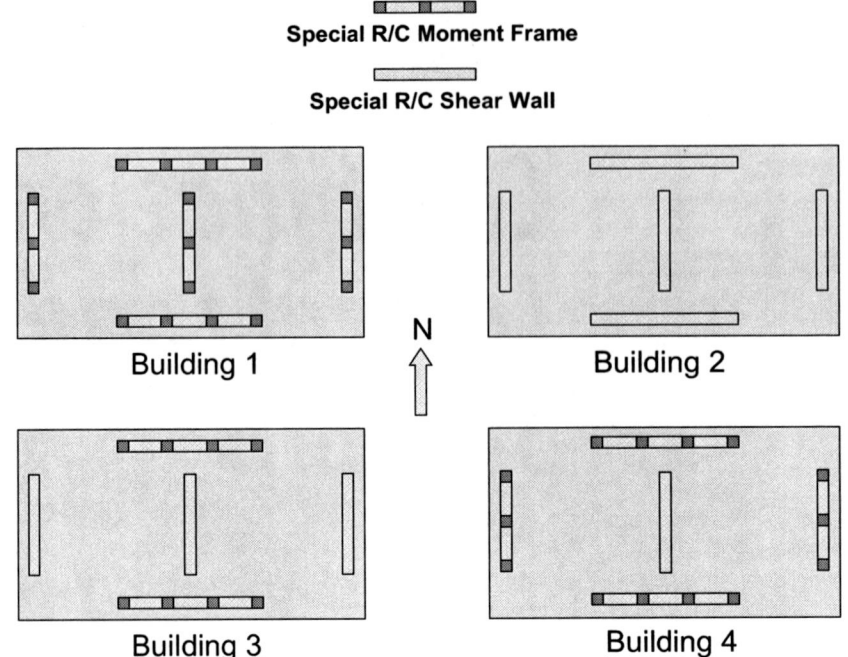

Figure G8-1 Planwise Combinations of Structural Systems.

the buildings is assigned to Seismic Design Category D. The shear wall systems are considered building frame systems and not bearing wall systems.

Buildings 1 and 2 have only one lateral system, and the system limitations and design parameters are taken directly from Table 12.2-1:

8.1.1 Building 1: Special RC Moment Frame in Both Directions

Height limit for SDC D = None
$R = 8$
$\Omega_o = 3$
$C_d = 5.5$

8.1.2 Building 2: Special RC Shear Wall in Both Directions

Height limit for SDC D = 160 ft
$R = 6$
$\Omega_o = 2.5$
$C_d = 5.0$

Building 3 has different systems in the two different directions, but there is only one system in each individual direction. Clearly, the height limitation for the building is controlled by the shear wall. The design values for each system are as specified for that system in Table 12.2-1. The values for Building 3 are as follows:

8.1.3 Building 3: Special RC Moment Frame in the East–West Direction and Special RC Shear Wall in the North–South Direction

Height Limit = 160 ft

For Special RC Moment Frame in the East–West Direction

$R = 8$
$\Omega_o = 3$
$C_d = 5.5$

For Special RC Shear Wall in the North–South Direction

$R = 6$
$\Omega_o = 2.5$
$C_d = 5.0$

Building 4 has a single system in the east–west direction and a combination of systems in the north–south direction. Assuming that the combined system is not designed as a dual system, Section 12.2.3 states that the more stringent system height limitation must be used in the north–south direction, hence the shear wall would control, and the height limitation for the structure would be 160 ft.

The design parameters for the moment frames acting in the east–west direction would be taken directly from Table 12.2-1. For the north–south direction, Section 12.2.3.2 states that the value of R used for the combined system would be the least value of R for any system used in the given direction. For the structure under north–south loading, the shear wall has the lowest R, and thus $R = 6$ is assigned. Section 12.2.3.2 further stipulates that the C_d and Ω_o values for the combined system would be taken from the system that governs R, thus again, the shear wall values control. The design values for Building 4 are summarized as follows:

8.1.4 Building 4: Special RC Moment Frame in the East–West Direction, Combination of Special RC Moment Frame and Special RC Shear Wall in the North–South Direction. Not Designed as a Dual System in the North–South Direction

Height limit = 160 ft

For the Moment Frames in the East–West Direction

$R = 8$
$\Omega_o = 3$
$C_d = 5.5$

For the Combined System Acting in the North–South Direction

$R = 6$
$\Omega_o = 2.5$
$C_d = 5.0$

It would probably be beneficial to design the combined moment frame–shear wall as a dual system. The dual system has no height limitation, and the R value is 7, compared to that for the combined (non-dual) system, which has $R = 6$. However, the moment frame in the dual system must be designed to resist at least 25 percent of the design base shear. The parameters for Building 4 with the north–south direction designed as a dual system are as follows:

8.1.5 Building 4: Special RC Moment Frame in the East–West Direction, Combination of Special RC Moment Frame and Special RC Shear Wall in the North–South Direction. Designed as a Dual System in the North–South Direction.

Height limit = None

For the Moment Frames in the East–West Direction

$R = 8$
$\Omega_o = 3$
$C_d = 5.5$

For the Dual System Acting in the North–South Direction

$R = 7$
$\Omega_o = 2.5$
$C_d = 5.5$

8.2 Combinations of Structural Systems in the Vertical Direction

Section 12.2.3.1 provides the requirements for systems with different systems in the vertical direction. This kind of system is shown as Buildings C and D in **Fig. G8–2**. Building C has an X-braced system on the bottom six levels and a moment frame in the top six levels. Building D in the same figure is just the opposite, with the moment frame at the bottom and the braced frame at the top. Buildings A and B of **Fig. G8–2** consist of a moment frame or a braced frame for the full height. In all cases, the moment frame is a special steel moment resisting frame, and the braced frame is a special steel concentrically braced frame. The height of each of the buildings is 150 ft, and the buildings are assigned to Seismic Design Category C.

Buildings A and B have a single system along the full height, and their limitations and design parameters come directly from Table 12.2-1. The system values are summarized for these two buildings as follows:

8.2.1 Building A: Special Steel Moment Frame

Height limit for SDC C = None
$R = 8$
$\Omega_o = 3$
$C_d = 5.5$

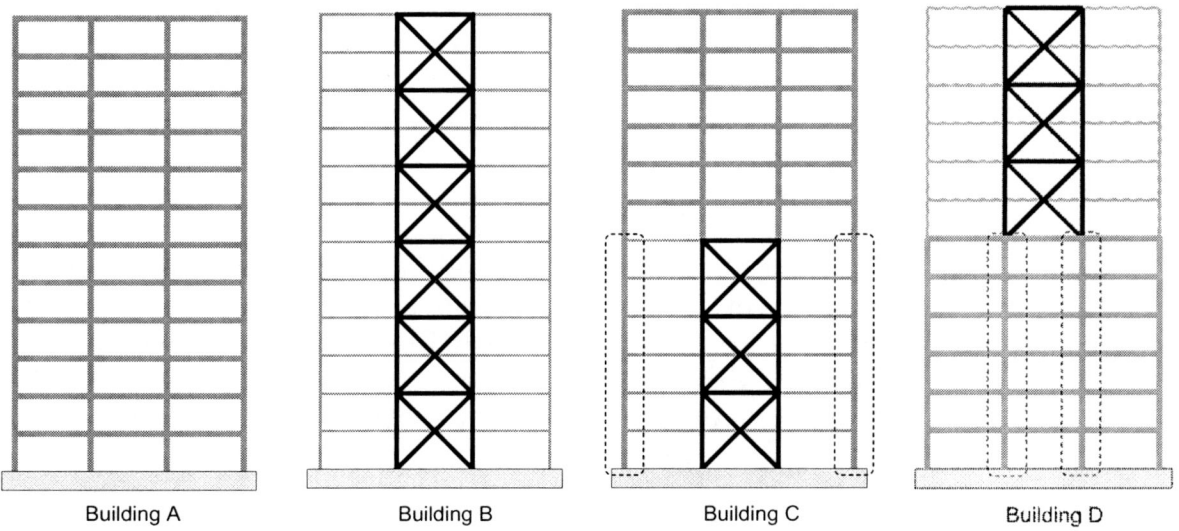

Building A Building B Building C Building D

Figure G8–2 Heightwise Combinations of Structural Systems.

8.2.2 Building B: Special Steel Concentrically Braced Frame

Height limit for SDC C = None
$R = 6$
$\Omega_o = 2.0$
$C_d = 5.0$

Building C has a special steel concentrically braced frame (CBF) in the bottom six stories and a special steel moment frame in the top six stories. Section 12.2.3.1 provides the requirements for selecting the system limitations and design parameters. This section states that the value for R used for any story shall not exceed the lowest value of R used for any story above that story. Likewise, the value of Ω_o and C_d used for any story shall not be less than the largest value used for any story above. Thus, for this building, $R = 8$, $\Omega_o = 3$, and $C_d = 5.5$ can be used for the upper six levels, and $R = 6$, $\Omega_o = 3.0$, and $C_d = 5.5$ must be used for the lower stories.

Section 12.2.3.1 does not state explicitly which height limitation would apply, but this situation is irrelevant for the given example because each system has a height limit that is greater than the height of the building. In cases where there is a potential for the height limit for one of the components of a combined system to be less than the intended height of the building, Section 12.2.3 indicates that the more stringent height limitation would govern.

The design values for Building C are as follows:

8.2.3 Building C: Stories 1 Through 6 (Special Steel Concentrically Braced Frame)

$R = 6$
$\Omega_o = 3.0$
$C_d = 5.5$

8.2.4 Building C: Stories 7 Through 12 (Special Steel Moment Frame)

$R = 8$
$\Omega_o = 3.0$
$C_d = 5.5$

There could be some question as to how the exterior columns in the lower six stories of Building C would be designed. These columns are shown within the dotted line regions of **Fig. G8–2** (Building C). Strictly speaking, these columns are part of the special moment frame because they transmit the overturning moment of the upper six levels to the base of the building. These columns should be detailed as special moment frame columns but proportioned on the basis of $R = 6$. Additionally, on the basis of the requirements of Section 12.3.3.3, it may be argued that these columns should be designed for forces based on load combinations that include the overstrength factor Ω_o.

Building D has a special moment frame in the lower six stories and special steel CBF in the upper six stories. According to Section 12.2.3.1, the lower stories should use the lower value of R as determined from the upper stories and the higher value of Ω_o and C_d as determined from the lower stories. Hence, the design values for the structure are as follows:

8.2.5 Building D: Special Steel Moment Frame for Stories 1 Through 6

$R = 6$
$\Omega_o = 3.0$
$C_d = 5.5$

8.2.6 Building D: Special Steel Concentrically Braced Frame for Stories 7 Through 12

$R = 6$
$\Omega_o = 2.0$
$C_d = 5.0$

As with Building C, a question arises with regard to the design of the discontinuous columns in Building D. Here, the interior columns are at issue. These columns are enclosed by dotted lines in **Fig. G8–2** (Building D). Section 12.3.3.3 requires that elements supporting discontinuous frames be designed using load factors that include the overstrength factor Ω_o when a Type 4 vertical irregularity (in-plane discontinuity irregularity) exists. It is not clear whether an irregularity exists because there is no apparent offset in the lateral load resisting system and because it is not clear whether there is a reduction in stiffness below the transition. However, some engineers might consider the offset in the lateral load resisting system to be infinite, thereby requiring the design of the interior columns with the factor Ω_o. (The author would not be inclined to use the Ω_o factors for the exterior columns of Building C but would use the Ω_o factors for the interior columns of Building D.)

As a final point, ASCE 7 places severe restrictions on systems similar to Building C in **Fig. G8–2** (moment frames above braced frames) when such systems are used in buildings assigned to SDC D and above. These restrictions, given in Section 12.2.5.5, are not applicable to the buildings in this example because the SDC was C.

8.3 Computing Approximate Periods of Vibration for Combined Systems

It is almost always necessary to determine the approximate period of vibration, T_a, for a structure. This period is used in computing the seismic base shear when the equivalent lateral force (ELF) method of analysis is used and for scaling the results of a modal response spectrum analysis when the base shear from such an analysis is less than 85 percent of the ELF base shear (Section 12.9.4). The approximate period is computed using Eq. 12.8-7, which uses the parameters C_t and x. Table 12.8-2 provides these parameters for a variety of well-defined systems but does not specifically address combined or dual systems. Hence, dual systems or combined systems would apparently fall under the "All other structural systems" category. This approach seems overly conservative, and it would seem reasonable to use a weighted average based on the parameters in Table 12.8-2. For example, the period for Building D of **Fig. G8–2** could be estimated as the average of the period of a 12-story moment frame and a 12-story braced frame. However, the average of the periods of a 6-story moment frame (the top half) and a 6-story braced frame (the bottom half) would not be appropriate. Nor would it be appropriate to use one period for the upper half of the building and a different period for the lower half.

8.3.1 Performing Structural Analysis for Combined Systems

When designing buildings with mixed R values, it is recommended that the analysis be performed with $R = 1$ and that the actual R values be assigned on a member-by-member basis. In this case, interstory drifts would be multiplied by C_d/R, where again, the actual R value would be used.

8.3.2 Vertical Combination When the Lower Section Is Stiff Relative to the Upper Portion

Where the lower portion of a building is much stiffer than the upper portion, it is permitted to use a two-stage ELF procedure to determine design forces. Section 12.2.3.1 stipulates that the two-stage analysis is limited to systems in which the lower portion is at least 10 times as stiff as the upper portion and for which the period of the entire structure is not greater than 1.1 times the period of the upper portion of the structure, with the upper portion fixed at its base.

The stiffness requirements are difficult to apply because the standard does not specify how the stiffness is to be computed, and there can be several measures of the stiffness of a structure. Because of these complexities, examples of this type of system are deferred to Example 18, which covers the equivalent lateral force method.

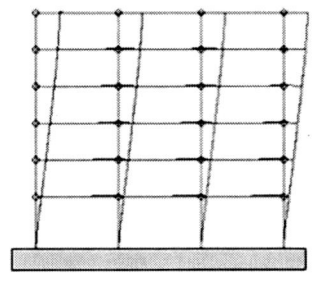

Example 9
Horizontal Structural Irregularities

Section 12.3.2.1 is used to determine if one or more horizontal structural irregularities exist in the lateral load resisting system. The five basic irregularity types are described in Table 12.3-1. This example explores each of these irregularities but concentrates primarily on horizontal structural irregularity Types 1a and 1b (torsion and extreme torsion).

9.1 Torsional Irregularities (Types 1a and 1b)

Based on the definitions in Table 12.3-1, a torsional irregularity exists if the story drift at one end of a building is more than 1.2 times the average of the drift at the two ends of the building when that building is subjected to lateral forces applied at the minimum eccentricity of 0.05 times the length of the building between the two ends. If the ratio of story drifts is greater than 1.4, an extreme torsional irregularity exists. Torsional irregularity must only be checked for systems with rigid or semirigid diaphragms.

Two examples are provided for determining if a torsional irregularity exists. In the first example, a simple one-story structure with symmetrically placed shear walls is analyzed to determine the effect of the placement of the walls on the torsional stiffness of the system. The second example is more realistic in the sense that it determines whether torsional irregularities occur in a typical four-story office building.

9.1.1 Torsion in a Simple One-Story Structure

Consider first the simple one-story system shown in **Fig. G9–1**. The lateral system for the building consists of four walls, each with lateral stiffness k. The walls are placed symmetrically in the system; the parameter α is used to locate the walls some distance from the center of mass, which is located at the geometric center of the building. The diaphragm is assumed to be rigid. The lateral load V is applied at an accidental eccentricity of 0.05 times the building width, L.

Given this configuration, the displacement at the center of the building is

$$\Delta_{CENTER} = \frac{V}{2k} \tag{G9-1}$$

and the deflection at the edge of the building is

$$\Delta_{EDGE} = \frac{V}{2k} + \frac{0.05VL}{2k(\alpha^2 L^2 + \alpha^2 B^2)}(0.5L) \tag{G9-2}$$

The deflection at the center of the building is the same as the average of the deflections at the extreme edges of this rigid diaphragm building.

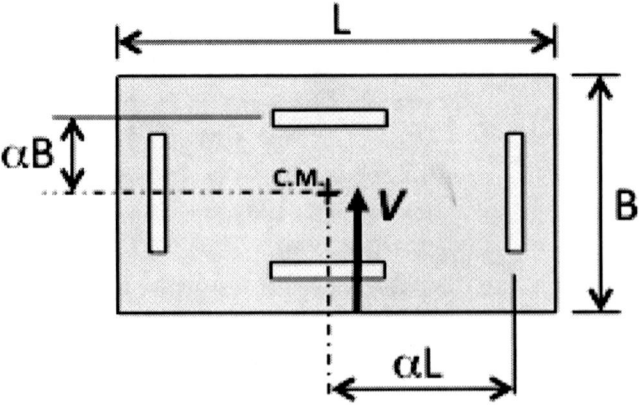

Figure G9–1 Simple Building for Accidental Torsion Evaluation.

Using Eqs. G9-1 and G9-2 and simplifying, the ratio of the displacement at the edge to the center is

$$Ratio = \frac{\Delta_{EDGE}}{\Delta_{CENTER}} = 1 + \frac{0.025}{\alpha^2 \left[1 + \left(\frac{B}{L}\right)^2\right]} \qquad (G9\text{-}3)$$

Fig. G9–2 is a plot of Eq. G9-3 for four values of B/L and for α values ranging from 0.1 (all walls near the center of the building) to 0.5 (all walls on the perimeter of the building). Also shown (using dashed lines) are the limits for a torsional irregularity (1.2) and for an extreme torsional irregularity (1.4). For the rectangular building with $B/L = 0.25$, the torsional irregularity occurs when α is approximately 0.35 and the extreme irregularity occurs when $\alpha = 0.24$. For the square building ($B/L = 1$), the torsional and extreme torsional irregularities occur at $\alpha = 0.25$ and 0.17, respectively.

Three important observations may be drawn from the results:

1. Torsional irregularities may occur even when the lateral load resisting system is completely symmetric.
2. The closer the walls are to the center of the building (lower values of α), the greater the possibility of encountering a torsional irregularity.
3. Torsional irregularities are more likely to occur in rectangular buildings than in square buildings.

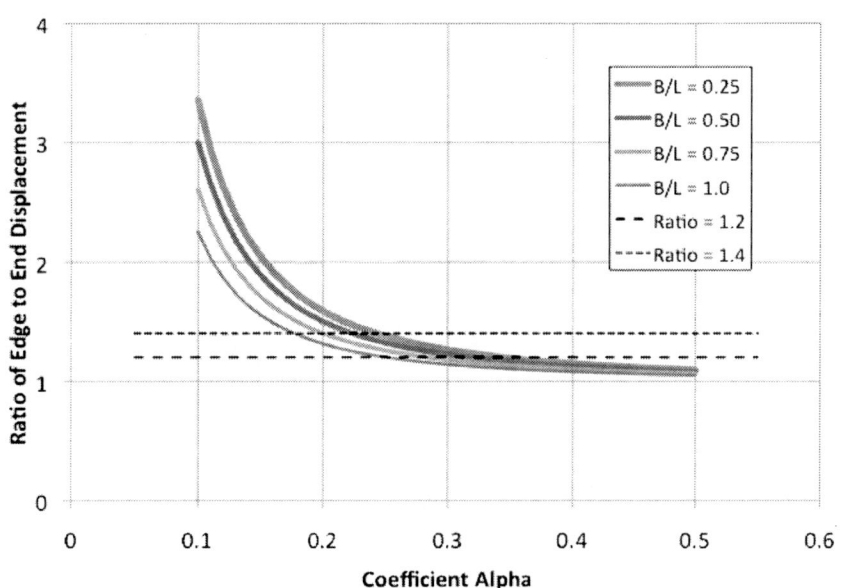

Figure G9–2 Effect of Wall Placement on Torsional Irregularity.

Section 12.8.4.3 of ASCE 7 requires that the accidental torsion be amplified in Seismic Design Categories C, D, E, and F. The amplification factor A_x is given as follows:

$$A_x = \left(\frac{\delta_{max}}{1.2\delta_{avg}}\right)^2 \leq 3.0 \qquad (12.8\text{-}14)$$

Fig. G9–3 is a plot of the amplification factors for the same four B/L ratios discussed above. Also shown on the plot is a dashed line with a constant value of 3.0, which represents the maximum amplification factor that must be used in analysis. For the rectangular building with $B/L = 0.25$, the maximum amplification occurs when α is 0.15, and for the square building, the maximum occurs when α is 0.11.

9.1.2 Example for Six-Story Building

The second example for checking if a torsional irregularity occurs is based on a six-story building with a typical floor plan as shown in **Fig. G9–4**. The lateral load resisting systems consist of moment frames on grid lines A and D and braced frames on lines 2, 3, 4, and 5. There is no bracing on grid line 6, so the center of mass and the center of rigidity do not coincide with respect to loading in the transverse direction. The height of the first story is 16 ft, and the height of stories 2 through 6 is 13 ft, giving a total building height of 81 ft.

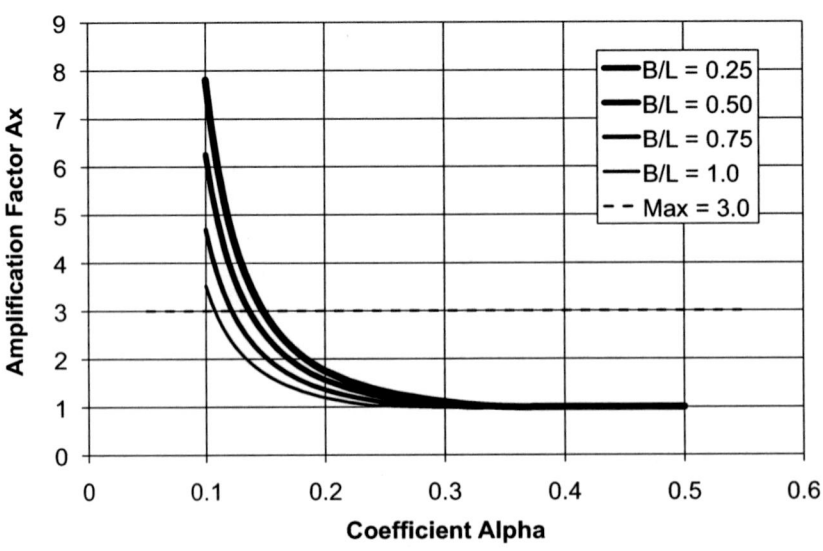

Figure G9–3 Effect of Wall Placement on Torsional Amplification.

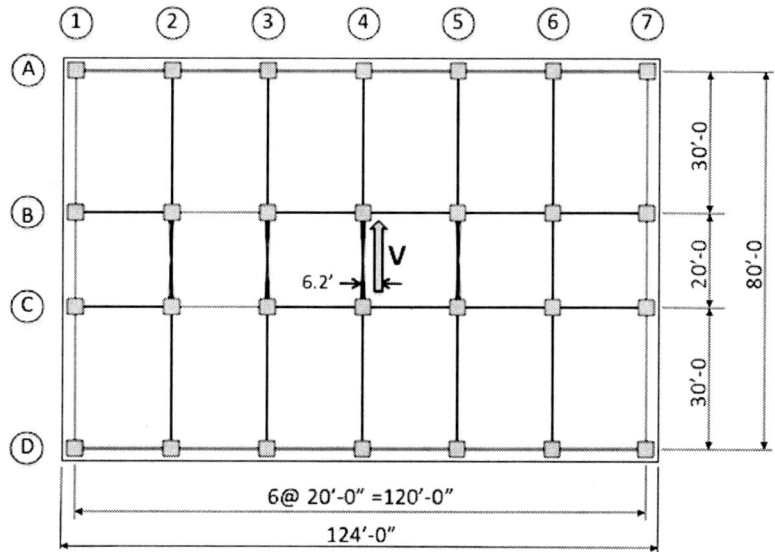

Figure G9–4 Plan View of Six-Story Building for Evaluation of Torsional Irregularity.

The results of the torsional analysis for loading in the transverse direction are shown in **Table G9–1**. The analysis was run using a rigid diaphragm assumption. Column 2 of the table shows the equivalent lateral forces at each level computed according to the equivalent lateral force procedure of Section 12.8. Column 3 of the table provides the accidental torsions at each level, which are equal to the story force times 6.2 ft (0.05 times the width of the building). The torsions are based on the lateral forces applied to the right of frame line 4 because this produces the largest displacements at the edge of the building. (In a more complicated building, this is not apparent, and both directions for the eccentricity should be checked.) Column 4 of the table provides the computed story displacements at grid line 4, and column 5 lists the interstory drifts on grid line 4. Columns 6 and 7 contain the displacements and interstory drifts at the edge of the building (2.0 ft to the right of grid line 7. The ratio of the drift at the edge of the building to the drift at the center is shown in column 8). As may be observed in column 8, the ratio of edge drift to center drift exceeds 1.2, but is less than 1.4, for stories 1 through 4, so the structure has a Type 1a torsional irregularity.

The torsional amplification factors, computed according to Eq. 12.8-14, are listed in column 9 of **Table G9–1**. These quantities are based on story displacements, not interstory drifts. For level 3, for example,

$$A_x = \left(\frac{\delta_{max}}{1.2\delta_{avg}}\right)^2 = \left(\frac{1.864}{1.2(1.512)}\right)^2 = 1.055$$

Example 14 of this guide contains additional discussion on accidental torsion and torsional amplification.

Table G9-1 Torsional Irregularity Check for the Building in **Fig. G9-4**

1	2	3	4	5	6	7	8	9
Level	F_i (kips)	T_{ai} (kips-ft)	δ_i Line 4 (in.)	Δ_i Line 4 (in.)	δ_i Line 7 (in.)	Δ_i Line 7 (in.)	Ratio (7/4)	Calculated A_x
6	207.0	1283.3	3.219	0.601	3.863	0.685	1.139	1.000
5	160.1	992.6	2.618	0.605	3.178	0.707	1.168	1.023[a]
4	111.9	693.5	2.013	0.501	2.471	0.607	1.212	1.047
3	70.9	439.7	1.512	0.472	1.864	0.583	1.235	1.056
2	37.9	235.0	1.040	0.464	1.281	0.588	1.268	1.054
1	15.0	93.0	0.576	0.576	0.693	0.693	1.203	1.005

a. An amplification factor of 1.0 could be applied at this level because there is no torsional irregularity

9.2 Reentrant Corner Irregularity (Type 2)

According to Table 12.3-1, a reentrant corner irregularity occurs when both plan projections are greater than 15 percent of the width of the plan dimension in the direction of the projection. **Fig. G9–5** shows the plan of a building with four reentrant corners, marked A through D. For this building, only corner D would cause a reentrant corner irregularity because both projections are greater than 15 percent of the width of the building.

In some cases, a notch in the edge of the building may trigger a reentrant corner irregularity. See the following discussion on diaphragm irregularities for more details.

9.3 Diaphragm Discontinuity Irregularity (Type 3)

According to Table 12.3-1, diaphragm discontinuity irregularities occur if the area of a notch or hole in the diaphragm is greater than 50 percent of the gross enclosed area of the diaphragm, or if the in-plane stiffness of the diaphragm at one level is less than 50 percent of the stiffness at an adjacent level.

Fig. G9–6 shows four different diaphragms. In part A of the figure, the opening is a notch and has less than 50 percent of the gross enclosed area, so the diaphragm is not irregular. Note the presence of the exterior window wall, which is required if the shown opening is to be considered as part of the enclosed area. If this window wall did not exist, the opening might in fact cause a reentrant corner irregularity because the projections caused by the opening are greater than 15 percent of the building width. This situation is shown in **Fig. G9–7**.

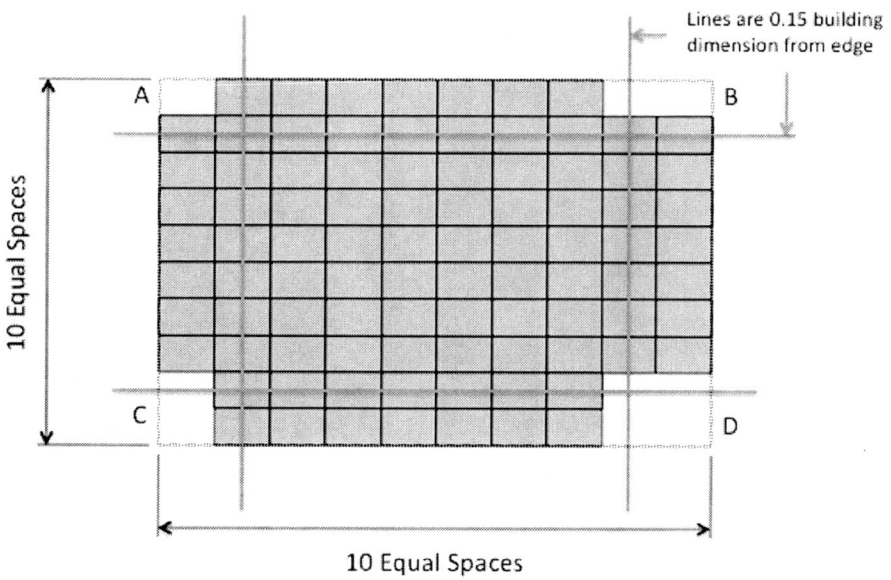

Figure G9–5 Building with Four Reentrant Corners and with a Reentrant Irregularity.

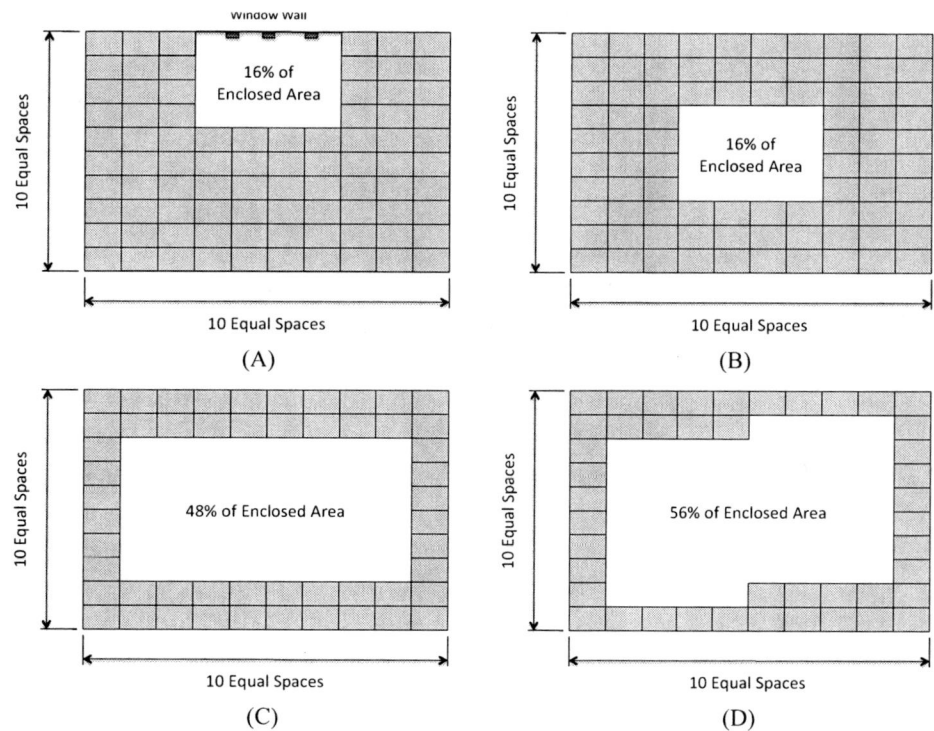

Figure G9–6 Diaphragm Openings and Irregularities.

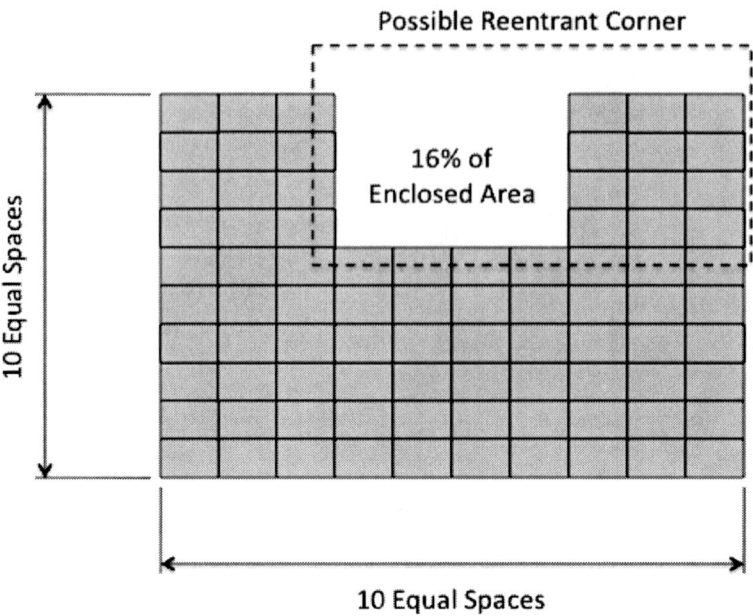

Figure G9–7 A Diaphragm Notch Causing a Reentrant Corner Irregularity.

In parts B and C of **Fig. G9–6**, the diaphragm openings are at the interior of the building, and neither triggers a diaphragm irregularity because the areas of the openings are less than 50 percent of the gross enclosed area. The opening in part D of the figure does cause a diaphragm irregularity.

A diaphragm irregularity can also occur if the in-plane stiffness at one level is less than 50 percent of the stiffness at an adjacent level. Calculations to determine diaphragm stiffness are not straightforward but in some cases can be accomplished using finite element analysis. For example, a finite element model for diaphragm B in **Fig. G9–6** is shown in **Fig. G9–8**. The diaphragm stiffness would be computed as the quantity V/Δ. In the figure, the lines to the left and right edges represent lines of support provided by lateral load resisting elements. For more complex systems, with multiple diaphragm segments and multiple lateral load resisting elements, the determination of diaphragm stiffness is essentially impossible in terms of the definition provided in Table 12.3-1.

A diaphragm irregularity based on a differing stiffness of adjacent stories may in fact be irrelevant. Consider, for example, a rectangular building with no diaphragm openings. At one level, the floor slab is 4 in. thick, and at an adjacent level, the thickness is 10 in. Clearly the 10-in.-thick diaphragm is more than twice as stiff as the 4-in.-thick diaphragm, but in terms of stiffness relative to the lateral load resisting system, both may be considered rigid. Hence, the different stiffness of the diaphragms has virtually no effect on the analysis, with the exception that the increased thickness may cause a mass irregularity. However, regardless of the actual diaphragm

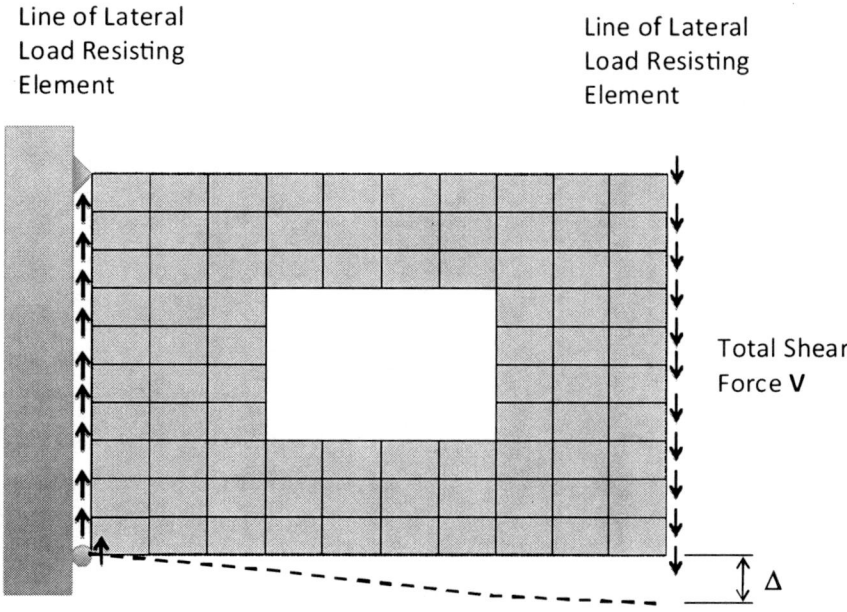

Figure G9–8 Finite Element Model for Determining Diaphragm Stiffness.

behavior, the classification of the diaphragm as discontinuous cannot be ignored when addressing the consequences of the irregularity (e.g., increased collector forces required by Section 12.3.3.4).

9.4 Out-of-Plane Offset Irregularity (Type 4)

Out-of-plane irregularities occur when the lateral forces in a lateral load resisting element are transferred to an element that is not in the same plane as that element. An example is shown in **Fig. G9–9** which is a plan view of the Imperial County Services Building, which was severely damaged during the October 15, 1979, Imperial Valley earthquake. In this building, wall A, an exterior wall, occurs on stories 2 through 6 and transfers shear to wall B, which exists only on the first story. Overturning moment is transferred to the columns adjacent to but offset from wall A. The walls labeled C extend the full height of the building. The lateral system in the long direction consists of moment resisting frames. During the earthquake, the columns adjacent to wall A but not on the same line as wall A failed because of the combined effect of overturning in the transverse direction and moment frame action in the longitudinal direction. A photograph of the building is shown in **Fig. G9–10**.

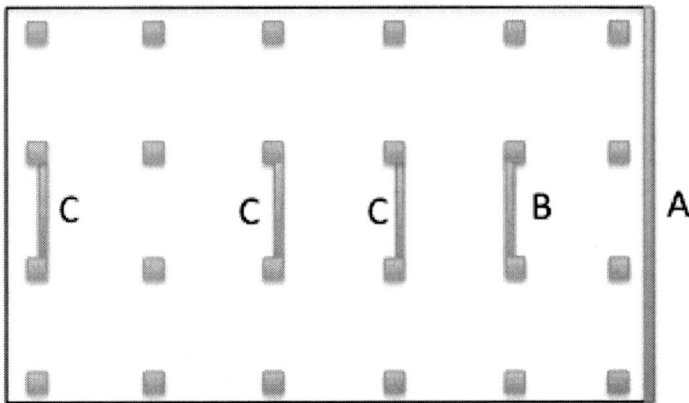

A) 2nd story and above shear wall
B) 1st story only shear wall
C) Full height shear wall

Figure G9–9   Plan View of Building with an Out-of-Plane Offset Irregularity.

Figure G9–10   Photograph of the Imperial County Services Building. Courtesy of V. Bertero.

9.5 Nonparallel System Irregularity (Type 5)

Nonparallel system irregularities occur when any element of the lateral load resisting system is not parallel to one of the orthogonal axes of the lateral load resisting system of the entire structure. Such a system is shown in **Fig. G9–11**, which is a plan view of a reinforced concrete frame-wall system. On the figure, the axes marked X and Y represent the principal axes of the

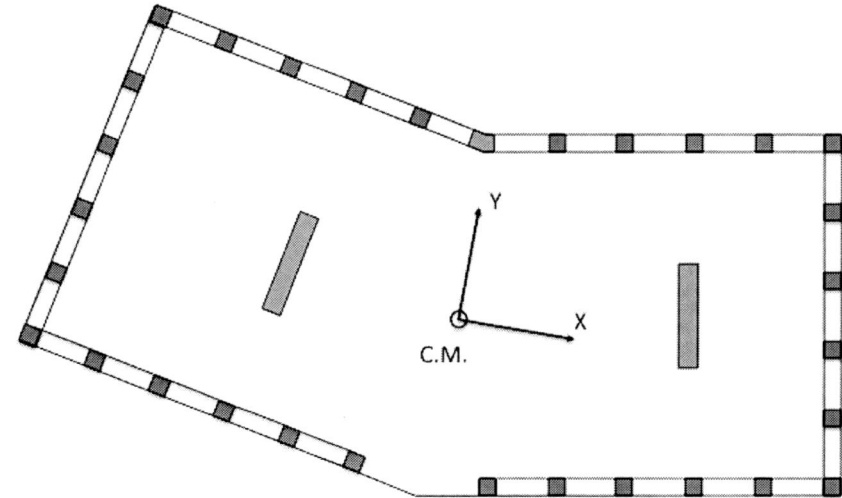

Figure G9–11 A Structure with a Nonparallel System Irregularity.

entire structural system. Clearly, the line of action of the lateral load resisting elements are not parallel to either the X or the Y axis, so a nonparallel system irregularity occurs.

When the lateral load resisting elements are parallel to but not symmetric about the system axes, the intent of Table 12.3-1 is that a nonparallel system irregularity exists. Hence, the structure shown in **Fig. G9–9**, with a nonsymmetric placement of walls, has a nonparallel system irregularity.

9.6 Consequences of Horizontal Irregularities

Horizontal irregularities are primarily of significance when the structure under consideration is assigned to SDC D or above, or in some cases, SDC C. The third and fourth columns of Table 12.3-1 provide the consequences of the irregularities in terms of the SDC. For example, a Type 5 horizontal irregularity triggers the requirement for consideration of orthogonal load effects (Section 12.5.3) in buildings assigned to SDC C and above and requires three-dimensional analysis in all SDC levels (Section 12.7.3). As seen from Table 12.3-2, vertical irregularities of any type influence the selection of the method of analysis (Table 12.6-1) for systems assigned to SDC C and above. Additionally, Section 12.3.3.1 provides circumstances in which certain horizontal irregularities are prohibited.

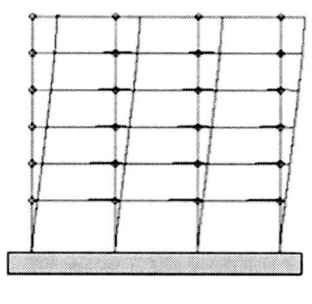

Example 10

Vertical Structural Irregularities

Section 12.3.2.2 is used to determine if one or more vertical structural irregularities exist in the lateral load resisting system. The five basic irregularity types are described in Table 12.3-2. This example explores each of these irregularities, but concentrates primarily on irregularity Types 1a and 1b (soft story), and Types 5a and 5b (weak story).

10.1 Soft Story (Stiffness) Irregularities (Types 1a and 1b)

The first step in a soft story irregularity check is to use the first exception in Section 12.3.2.2 to determine if there is potential for a soft story irregularity. This check is based on interstory drift and is not difficult to perform. If the relative drift criteria are met, there is no soft story irregularity and the check is complete. If the drift criteria are not met, the irregularity must be accepted or the stiffness-based check of Table 12.3-2 must be performed. The stiffness check has three possible results: no irregularity exists, a soft story irregularity exists, or an extreme soft story irregularity exists.

For the drift-based check, the structure is subjected to the design lateral loads, and interstory drift ratios are computed for each story. If the drift ratio in each story is less than 1.3 times the drift ratio in the story directly above it, there is no stiffness irregularity. When performing the drift check, the top two stories of the building need not be evaluated and accidental torsion need not be included.

Although ASCE 7 requires that the design-level lateral loads be used in the check, this is not strictly necessary when linear analysis is performed. It is important, though, that the vertical distribution of the lateral loads be reasonably correct. On the basis of this concept, the lateral loads used for the stiffness irregularity check may be based on the equivalent lateral force (ELF) method described in Section 12.8. The result of the calculation may show, however, that the ELF method may not be used for the final design of the structure. Applicability of the ELF method is covered in Section 12.6 and Table 12.6-1.

For this example, we consider a six-story reinforced concrete moment frame located in a region of high seismicity ($S_{D1} > 0.4\ g$). For simplicity, we assume that the story weight for each level is 1,500 kips and that all story heights are equal to 12.5 ft, except for one tall story, which has a height of 18.5 ft. Two separate analyses are run, one for which the tall story is the first story of the building, and one for which the tall story is the second story of the building. Elevations of these frames are shown in **Fig. G10–1**.

The period of vibration of the structure is estimated as

$$T = C_u T_a \qquad \text{(from Section 12.8.2)}$$

where C_u is taken from Table 12.8-1 and where

$$T_a = C_t h_n^x \qquad (12.8\text{-}7)$$

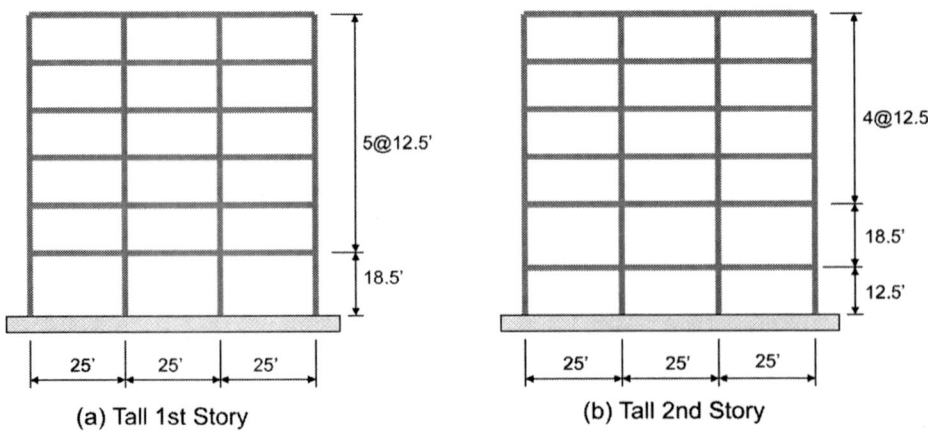

(a) Tall 1st Story   (b) Tall 2nd Story

Figure G10–1 Building Used to Investigate Soft Story Irregularity.

For our structure, h_n is 81 ft, and for a concrete moment frame, $C_t = 0.016$ and $x = 0.9$ (Table 12.8-2). Hence,

$$T_a = C_t h_n^x = 0.016 \times 81^{0.9} = 0.835 \text{ s}$$

$T = C_u T_a$ may only be used if a properly substantiated (computer) analysis has been used to determine the period. For this example, we assume that such an analysis has been performed and that the computed period exceeds $C_u T_a$, thereby setting $C_u T_a$ as the upper limit on period. This result produces a more realistic value of the exponent k than the use of T_a alone. Because of the high seismicity, $C_u = 1.4$ and

$$T_a = C_u T_a = 1.4 \times 0.835 = 1.17 \text{ s}$$

Based on the text in Section 12.8.3, the exponent k can be computed as

$k = 1$ when T is less than or equal to 0.5 s,
$k = 0.75\ T + 0.5$ when $0.5 < T < 2.5$ s, and
$k = 2$ when T is greater than or equal to 2.5 s

for the current example

$$k = 0.5\ T + 0.75 = 0.5(1.17) + 0.75 = 1.33$$

For computing the lateral force, a total base shear of 100 kips is assumed, and this shear is distributed vertically according to Eq. 12.8-12. **Table G10–1(a)** and **Table G10–2(a)** show the lateral load computations for the structures with the tall first and second story, respectively

The resulting story displacements, story drifts, story drift ratios, and ratio of story drift ratios are shown in **Table G10–1(b)** and **Table G10–2(b)**. These computations do not include the deflection amplification factor C_d because this factor cancels out when calculating the ratios of the story drift ratios.

Table G10–1a Development of Lateral Loads for Structure with Tall First Story

1	2	3	4	5	6	7
Story	H (ft)	h (ft)	W (kips)	Wh^k	Wh^k / Total	V (kips)
6	12.5	81.0	1,500	521,128	0.306	30.6
5	12.5	68.5	1,500	416,898	0.245	24.5
4	12.5	56.0	1,500	318,811	0.187	18.7
3	12.5	43.5	1,500	227,765	0.134	13.4
2	12.5	31.0	1,500	145,080	0.085	8.5
1	18.5	18.5	1,500	72,968	0.043	4.3
			Totals	1,702,650	1.000	100.0

Table G10–1b Drift-Based Soft-Story Check for Structure with Tall First Story

1	2	3	4	5	6
Story	H (in.)	δ (in.)	Δ (in.)	IDR	IDR_n/IDR_{n+1}
6	150	4.29	0.41	0.27	—
5	150	3.88	0.63	0.42	1.56
4	150	3.25	0.75	0.50	1.19
3	150	2.50	0.74	0.50	1.00
2	150	1.75	0.79	0.53	1.06
1	222	0.96	0.96	0.43	0.811

Table G10–2a Development of Lateral Loads for Structure with Tall Second Story

1	2	3	4	5	6	7
Story	H (ft)	h (ft)	W (kips)	Wh^k	Wh^k/Total	V (kips)
6	12.5	81.0	1,500	521,128	0.311	31.1
5	12.5	68.5	1,500	416,898	0.249	24.9
4	12.5	56.0	1,500	318,811	0.191	19.1
3	12.5	43.5	1,500	227,765	0.136	13.6
2	18.5	31.0	1,500	145,080	0.087	8.7
1	12.5	12.5	1,500	43,297	0.026	2.6
			Totals	1,672,979	1.000	100.0

Table G10–2b Drift-Based Soft-Story Check for Structure with Tall Second Story

1	2	3	4	5	6
Story	H (in.)	δ (in.)	Δ (in.)	IDR	IDR_n/IDR_{n+1}
6	150	4.60	0.42	0.28	—
5	150	4.18	0.65	0.43	1.54
4	150	3.54	0.78	0.52	1.21
3	150	2.75	0.82	0.55	0.96
2	222	1.93	1.47	0.66	1.20
1	150	0.46	0.46	0.31	0.47

In **Table G10–2(b)**, which is for the structure with the tall first story, the interstory drift ratio (IDR) in the tall bottom story (0.43 percent) is actually smaller than the drift ratio at the next story above (0.53 percent). The first-story drift ratio, divided by the second-story drift ratio is 0.811. This rather unexpected result occurs because the fixed-base condition stiffens the bottom story relative to the upper stories. This effect may be seen in the deflected shape profile, which is presented in **Fig. G10–2(a).**

The maximum ratio of IDRs in **Table G10–2(b)** is 1.56, which is for the fifth story relative to the sixth story. Although this ratio is greater than 1.3, it does not result in a soft story classification because the first exception in Section 12.3.2.2 states that the top two stories of the structure may be excluded from the check. By the nature of this exception, though, only buildings taller than two stories are subject to this irregularity.

The drift calculations are shown for the structure with the tall second story in **Table G10–2(b)**. Here, the second-story drift ratio (0.66 percent) is 1.2 times the drift ratio of the third story (0.55 percent). A soft story condition does not occur because this value is less than 1.3. The deflected shape profile for the structure with the soft second story is shown in **Fig. G10–2(b)**.

Based on the above results, both of the structures shown in **Fig. G10–1** are exempt from the stiffness-based soft story check described in the first two rows of Table-12.3-2. However, there are circumstances in which this check may be required. To illustrate this procedure, a stiffness-based soft story check is performed for the building with the tall second story (**Fig. G10–1(b)**), even though such a check is not actually required for this structure.

The first step in the analysis is the determination of the story stiffness. This determination is done on a story-by-story basis by applying equal and opposite lateral forces V at the top and bottom of the story, computing the interstory drift Δ in the story, and defining the interstory stiffness as

$$K_i = \frac{V}{\Delta_i}$$

After the story stiffnesses are determined, they are compared according to the requirements of Table 12.3-2. A soft story irregularity exists if for any story the stiffness of that story is less than 70 percent of the stiffness of the story above, or if the stiffness of the story is less than 80 percent of the average stiffnesses of the three stories above. The irregularities are considered extreme if for any story the stiffness of that story is less than 60 percent of the stiffness of the story above, or if the stiffness of the story is less than 70 percent of the average stiffnesses of the three stories above.

The results for the frame of **Fig. G10–1(b)** are shown in **Table G10–3**. A story shear of $V = 100$ kips was used in the analysis.

Based on this check, the structure has an extreme soft story irregularity. However, the structure need not be classified as such because the drift-based check of the same structure exempted this structure from the stiffness-based

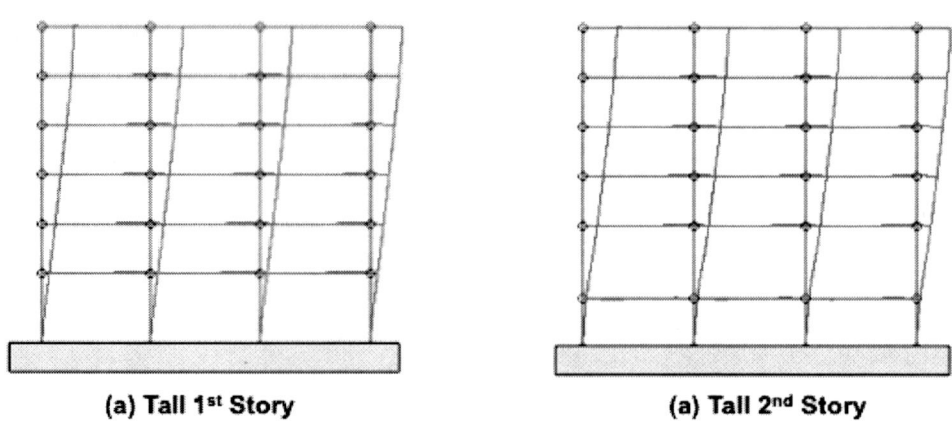

(a) Tall 1st Story   (a) Tall 2nd Story

Figure G10–2   Deflected Shape Profiles.

Table G10–3   Stiffness-Based Soft Story Analysis for Structure with Soft Second Story

1	2	3	4	5
Story	Δ (in.)	K (kips/in.)	K_n/K_{n+1}	$K_n/Avg\ K_{n+1}$
6	0.763	131	—	—
5	0.691	145	145/131 = 1.11	—
4	0.622	161	161/145 = 1.11	—
3	0.511	196	196/161 = 1.22	196/146 = 1.35
2	1.042	96	96/196 = 0.490	96/167 = 0.575
1	0.290	345	345/96 = 3.59	345/151 = 2.28

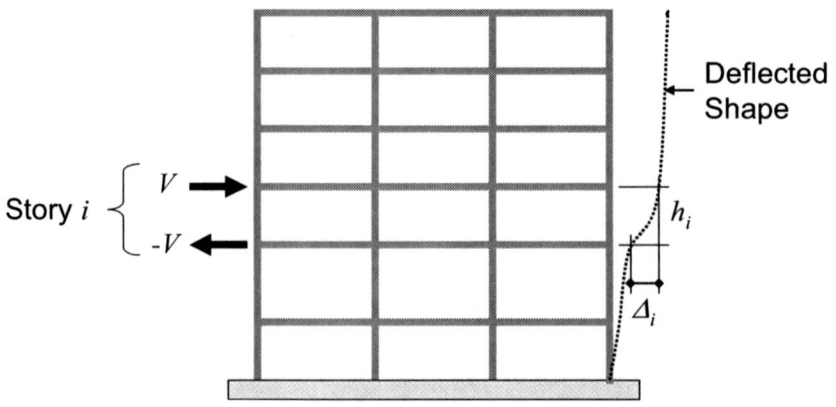

Figure G10–3   Computing Stiffness for Story i.

check. Based on this observation, one should always perform the drift-based check first because this step may exempt a structure from being classified as having a soft story irregularity, thereby allowing the designer to skip the more time-consuming stiffness-based check.

10.2 Weight (Mass) Irregularity (Type 2)

Vertical weight irregularities are relatively straightforward, and no example is presented. Note, however, that the story weight used in the calculation is the effective seismic weight, as defined in Section 12.7.2. Also, the same exception (the drift ratio test) that applies to stiffness irregularities may be applied to weight irregularities. It is possible that this exception may supersede the mass ratio test provided in Table 12.3-2

10.3 Vertical Geometric Irregularity (Type 3)

A vertical geometric irregularity occurs when the horizontal dimension of the lateral resisting system at one level is more than 130 percent of that for an adjacent story. Based on this definition, the structure shown in **Fig. G10–4(a)** has a geometric irregularity because the moment resisting frame has three bays on the third story and only two bays on the fourth story. On the other hand, the structure shown in **Fig. G10–4(b)** does not have a vertical geometric irregularity because the braced frame, which is the lateral resisting system, has the same horizontal dimension for the full height. The setbacks on the upper three stories have no influence on the vertical geometric irregularity of the braced frame system because the two exterior bays resist gravity loads only, and as such, are not part of the lateral load resisting system. It is possible (but unlikely given the configuration) that the moment frame also has a stiffness irregularity, and it is likely that both systems have a weight irregularity.

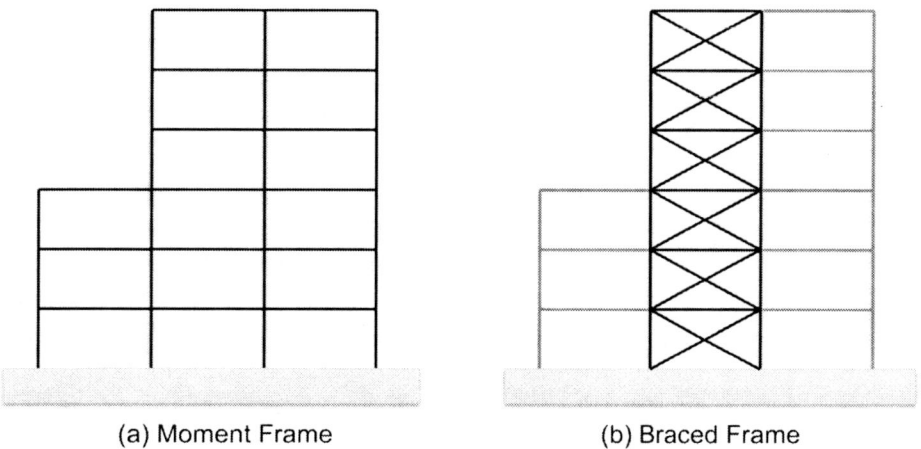

(a) Moment Frame (b) Braced Frame

Figure G10–4 Example of Vertical Geometric Irregularity.

10.4 In-Plane Discontinuity in Vertical Lateral Force-Resisting Element Irregularity (Type 4)

An in-plane discontinuity occurs when there is a horizontal offset in the lateral load resisting system and the distance of the offset is greater than the width of the lateral load resisting system. Based on this definition, the system shown in part (a) of **Fig. G10–5** has an irregularity because the offset dimension is two times the width of the lateral system. For the system shown in **Fig. G10–5(b)**, the system is on the cusp of having an in-plane discontinuity irregularity because the offset dimension is exactly equal to the width of the lateral load resisting system.

In system (a) shown in **Fig. G10–5**, the columns under the discontinuous braced frame (shown in the figure surrounded by a dotted line) would be subject to the requirement of Section 12.3.3.3, thereby requiring the columns to be designed with load cases that include the overstrength factor Ω_o. Strictly speaking, the column surrounded by the dotted line in part (b) of the figure would not be subject to the requirements of Section 12.3.3.3 because the offset is exactly equal to (not greater than) the width of the braced frame. However, in this situation, it is recommended that the column be designed using the overstrength factor Ω_o.

10.5 Discontinuity in Lateral Strength–Weak Story Irregularity (Types 5a and 5b)

Weak story irregularities are difficult to detect because the concept of story strength is not well defined, even for relatively simple systems, such as moment frames and X-braced frames. Providing a full numerical example of this type of irregularity is beyond the scope of this guide because the computation of story strength depends on rules established in the material specifications, such as ACI 318 (ACI 2005) or the AISC Seismic Provisions (AISC 2005a).

Some discussion is warranted, however. Consider the case of a moment resisting frame shown in **Fig. G10–6**. In part (a) of the figure, it is assumed that a column mechanism has formed. This kind of mechanism can form if the columns are weak relative to the beams (which is not allowed for special moment frames in steel or reinforced concrete). The mechanism shown in part (b) of the figure is a beam mechanism. This mechanism, as shown, is in fact incorrect because a single-story beam mechanism cannot occur because plastic hinges would need to form in all beams at all stories. Additionally, it would be necessary for hinges to form at the base of the columns if the columns are fixed at the base. (A mechanism could form with plastic hinges at both ends of all the beams at a given level if hinges also form in all the columns above and below that level.)

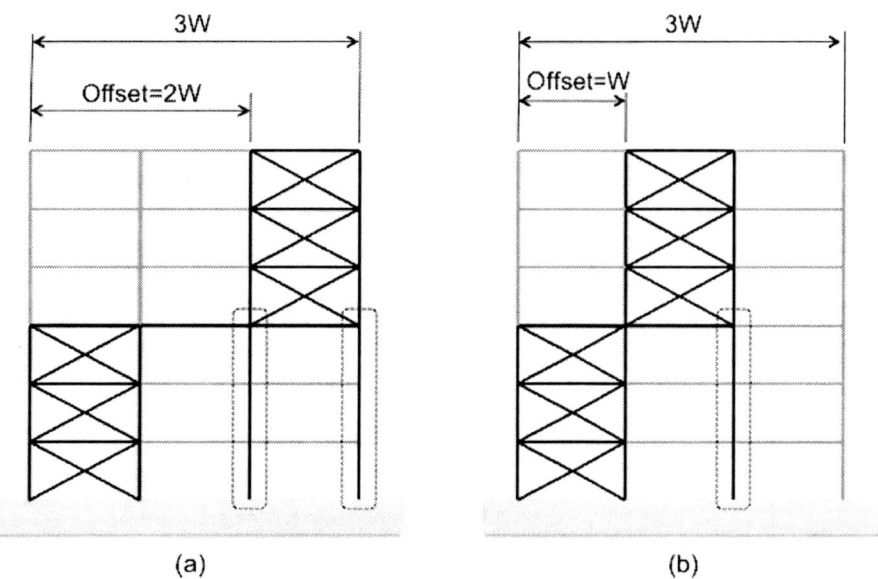

Figure G10–5 Example of an In-Plane Discontinuity Irregularity.

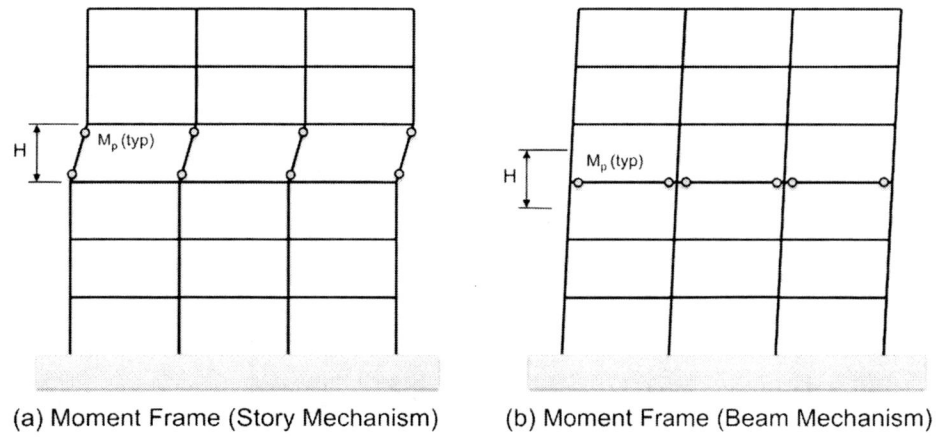

(a) Moment Frame (Story Mechanism)   (b) Moment Frame (Beam Mechanism)

Figure G10–6 Moment Frame Mechanisms.

On the basis of these assumptions, the commentary to the AISC Seismic Provisions provides formulas for computing the story strength for the mechanisms shown in **Fig. G10–6**. The intended use of the AISC expressions is related to frame stability, and not system irregularity.

For the story mechanism (**Figs. G10–6(a)**), the story strength of story, V_{yi}, is

$$V_{yi} = \frac{2\sum_{k=1}^{m} M_{pCk}}{H} \quad \text{(AISC Seismic Provisions C3-3)}$$

Seismic Loads: Guide to the Seismic Load Provisions ASCE 7-05 and similarly, for the beam (girder) mechanism, **Fig. G10–6(b)**,

$$V_{yi} = \frac{2\sum_{j=1}^{n} M_{pGj}}{H} \quad \text{(AISC Seismic Provisions C3-2)}$$

where

k and j are integer counters,

m is the number of columns,

M_{pCk} is the plastic moment strength of column k under minimum factored load,

n is the number of bays,

M_{pGj} is the plastic moment strength of beam j, and

H is the story height.

The above formulas would be equally applicable to structures of reinforced concrete and are suitable for computing story strength in association with the requirements of Table 12.3-2. For steel or concrete, the effect of the axial force in the columns on moment strength of the plastic hinges must be considered.

For braced frames, the story strength depends primarily on the bracing configuration, the axial strength of the brace, and the angle of attack θ of the brace, as shown in **Fig. G10–7**. In part (a) of the figure, the system is a buckling restrained braced frame, for which the strength of the single brace is the same in tension as it is in compression. For the concentrically braced frame (part (b) of **Fig. G10–7**), the strength of the tension and compression brace is different, and this difference has to be taken into consideration.

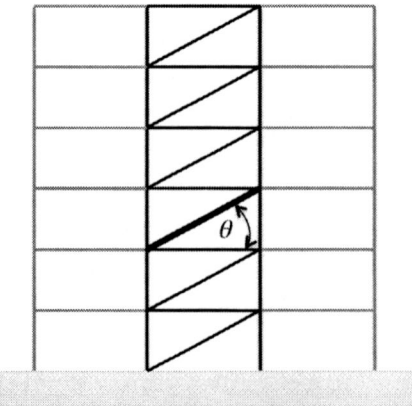
(a) Buckling Restrained Braced Frame

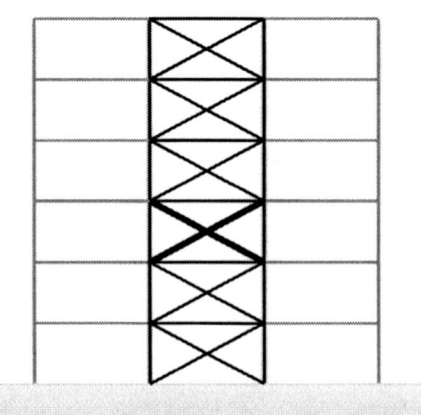
(b) Concentrically Braced Frame

Figure G10–7 Brace Frame Strength Irregularities.

Finding the story strength of other systems, such as frame-wall systems in concrete or moment frames in combination with braced frames in steel, is not straightforward and must be computed using nonlinear static analysis. A loading similar to that used to determine story stiffness (**Fig. G10–3**) could be used. Such analysis should include gravity load effects if it is expected that the gravity loads have an influence on member strength. A rigorous analysis would include P-delta effects as well.

Weak structure irregularities are highly undesirable and should be avoided if at all possible. Fortunately, such irregularities are uncommon in structures designed according to ASCE 7 and the material specifications (e.g., the AISC Seismic Provisions). The rarity of the irregularity is due to the fact that the design story shears always increase from the top to the bottom of the structure, and, hence, the story strengths should increase from the top to the bottom as well.

10.6 Consequences of Vertical Irregularities

Vertical irregularities are significant primarily when the structure under consideration is assigned to SDC D or above, or in some cases, SDC C. The second and third columns of Table 12.3-2 provide the consequences of the irregularities in terms of the seismic SDC. Section 12.3.3.1 prohibits weak story irregularities in SDC E or higher and prohibits extreme weak story irregularities in SDC D or higher.

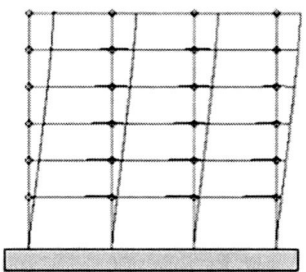

Example 11
Diaphragm Flexibility

Roof and floor diaphragms may be classified as flexible, semirigid, or rigid. Section 12.3.1.1 describes the conditions under which a diaphragm may be considered flexible, and Section 12.3.1.2 establishes the conditions under which the diaphragm may be considered rigid. If the diaphragm cannot be classified as flexible or rigid under these rules, an analytical procedure described in Section 12.3.1.3 is required. This procedure results in the classification of the diaphragm as either flexible or semirigid. This example demonstrates the analytical procedure. The descriptions under which a diaphragm is flexible according to Section 12.3.1.1 are not always accurate, and indeed, a flexible diaphragm does not really exist. Most diaphragms, even untopped metal deck or wood diaphragms with length-to-depth ratios up to

4.0, tend to be more rigid than flexible. In this context, ASCE 7 never requires a diaphragm to be modeled as flexible; however, in some cases it does permit it.

The structure to be considered in the example is shown in plan and elevation in **Figs. G11–1(a)** and **G11–1(b)**, respectively. The purpose of the structure is storage for hazardous chemicals. The lateral load resisting system in the transverse (Y) direction consists of four reinforced concrete shear walls, each 10 in. thick. The lateral load resisting system in the longitudinal (X) direction consists of 10-in. walls, 11 ft long, placed at the center of each bay. The diaphragm, also constructed from concrete, is 4 in. thick.

The length of the diaphragm between walls is 44 ft, and the depth of the diaphragm is 12 ft, producing a span-to-depth ratio of 3.67. According to Section 12.3.1.2, the diaphragm cannot automatically be considered rigid because the span-to-depth ratio is greater than 3.0[1]. An analysis must be performed to determine if the diaphragm is flexible or semirigid. According to Fig. 12.3-1, the diaphragm is flexible if the maximum diaphragm deflection (MDD) is greater than the 2.0 times the average drift of vertical element (ADVE).

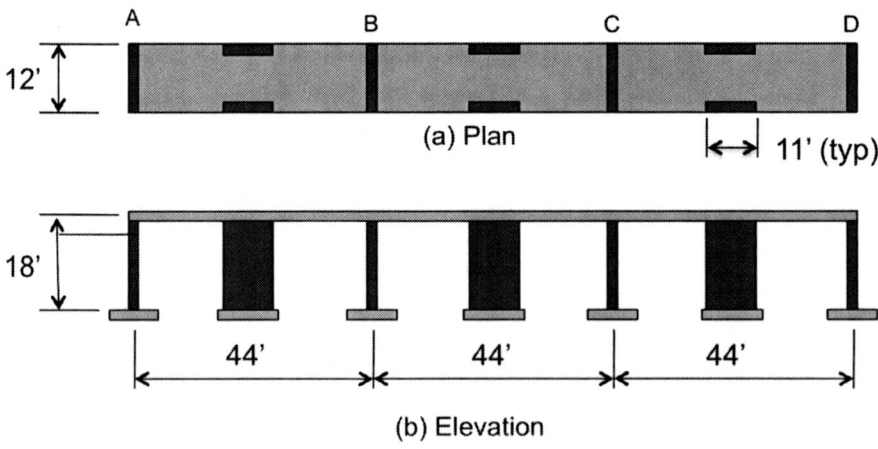

Figure G11–1 Three-Bay Concrete Structure Analyzed for Diaphragm Flexibility.

---

1. ASCE 7 is not clear on whether this ratio is related to the overall dimension of the diaphragm, or of a diaphragm segment (that portion of the diaphragm between two parallel lateral load resisting elements). This rather unrealistic example was devised to produce a span to depth ratio greater than 3.0 for a single diaphragm segment.

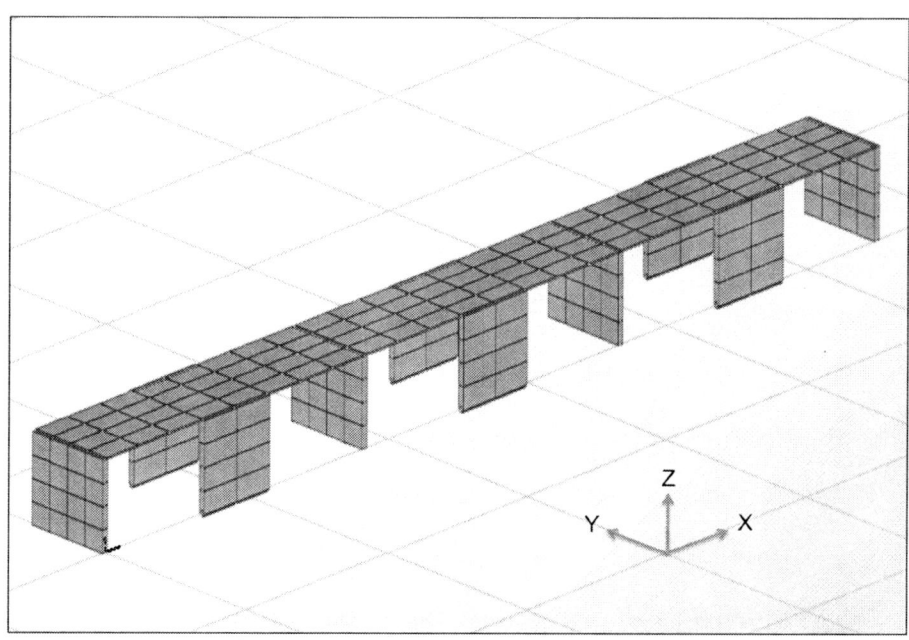

Figure G11-2 Finite Element Model for Computing Diaphragm Flexibility.

An analysis to determine the quantities *MDD* and *ADVE* was performed using the SAP2000 (CSI 2009) finite element analysis program. Thin-shell elements were used to model the walls and diaphragms. These elements automatically include in-plane shear deformations, which are essential for diaphragm analysis. An image of the finite element model is shown in **Fig. G11-2**. Loading consisted of a 10-kips force applied in the *Y* direction at each node along the edge of the diaphragm. A uniform load applied to the edge of the diaphragm may also be used and would be more appropriate where the shell elements are not of constant width. A distributed load has been used in lieu of a concentrated load (as implied by Fig. 12.3-1) because the inertial forces in a diaphragm would not be developed within a single point in an actual diaphragm. The use of a uniform load in lieu of a concentrated load also lowers the likelihood that the diaphragm would be classified as flexible.

The deflections computed along the edge of the diaphragm are shown in **Fig. G11-3**. In the end span, the average drift of the vertical element (Fig. 12.3-1) is

$ADVE = (0.0136 + 0.0295)/2 = 0.0216$ in.

The maximum diaphragm deflection, *MDD*, is

$MDD = 0.0425 - 0.0216 = 0.0209$ in.

The ratio of the maximum diaphragm deflection to the average displacement is

$MDD/ADVE = 0.0209/0.0216 = 0.967$

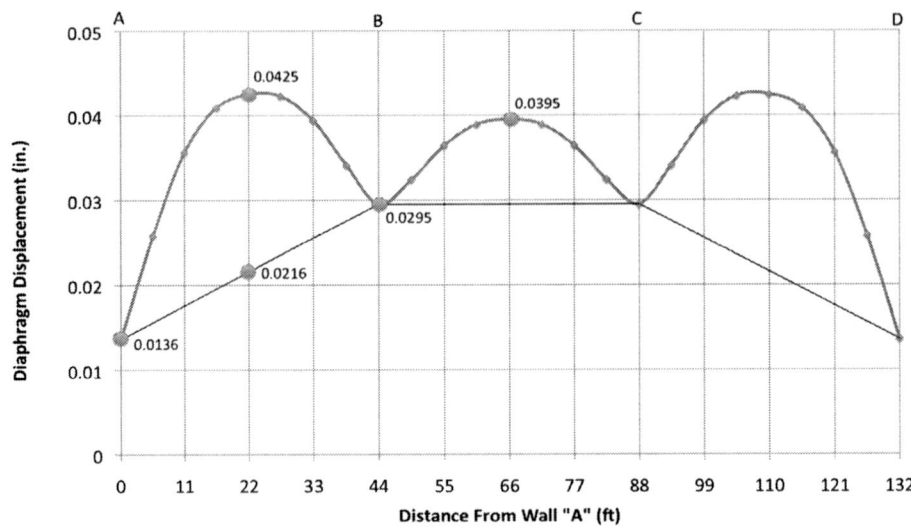

Figure G11–3 Diaphragm Displacements for a 4.0-in.-Thick Slab.

This ratio, 0.967, is less than 2.0, so according to Section 12.3.1.3, the diaphragm may not be considered flexible.

With a 10-kips force applied at each edge node, the total applied load on the structure is 250 kips. **Fig. G11–4** shows the distribution of these forces to the interior and exterior walls for a variety of assumptions. For the computed assumption, based on the finite element analysis, each of the exterior walls resists 38.4 kips, and the interior walls carry 83.0 kips each, for a total (all four walls) of 242.8 kips. This total is less than 250 kips because the transverse walls carry some shear, which is delivered to the foundation through weak-axis bending in these walls. For a fully rigid diaphragm, it would be expected that the 250-kips force would be equally distributed to the walls because the walls have the same lateral stiffness. This situation would result in a force of 62.5 kips in each wall. For a fully flexible diaphragm, the distribution of forces would be distributed on a tributary area basis with 1/6 of the total force, or 41.7 kips, going to the exterior walls, and 1/3 of the force, or 83.3 kips, going to each of the interior walls.

The finite element results appear to be more consistent with the flexible diaphragm assumption than with the rigid diaphragm assumption. The reason for this behavior actually has nothing to do with diaphragm flexibility. The behavior is based on the fact that the walls have virtually no torsional stiffness about the vertical axis, so these supports emulate more the condition of a pinned support than a fixed support in a three-span continuous beam. For example, a three-span continuous beam subjected to the same loading as that used in the finite element analysis would have exterior support reactions of 33.3 kips and interior reactions of 91.7 kips. This result, shown as the beam assumption in **Fig. G11–4**, is valid regardless of

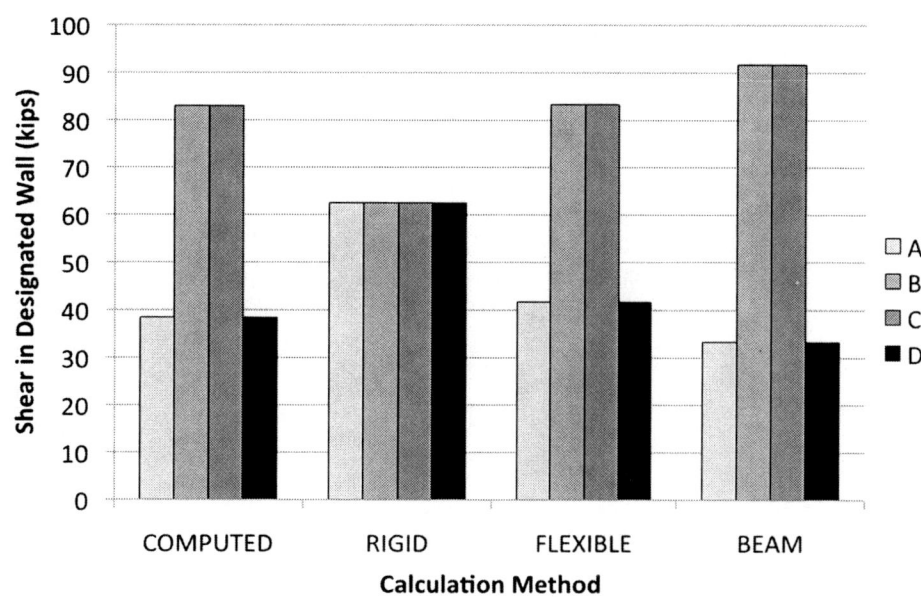

Figure G11–4 Distribution of Forces in Walls for Diaphragm Thickness of 4.0 in.

the stiffness of the beam, as long as each span has the same flexural stiffness. Therefore, a prudent designer should consider how sensitive the results are to the assumptions made. Other parameters, such as cracking in walls and diaphragms, can influence the force distribution in the lateral system. If the results are sensitive to various parameters, the design should be based on a bounded solution.

If we assume that the diaphragm is only 2.0 in. thick (e.g., concrete over a metal deck), the results of the finite analysis indicate that the diaphragm is semirigid. The deflected shape for this condition is provided in **Fig. G11–5**. The computations are as follows:

$$ADVE = (0.0134 + 0.0297)/2 = 0.0215 \text{ in.}$$

The maximum diaphragm deflection, MDD, is

$$MDD = 0.0595 - 0.0215 = 0.0380 \text{ in.}$$

The ratio of the maximum to the average displacement is

$$MDD/ADVE = 0.0380/0.0215 = 1.767$$

This value is less than 2.0, so even the system with the 2.0-in. slab is classified as semirigid.

The computed reactions at the base of the walls are close to those determined for the system with the 4.0-in.-thick diaphragm. Again, this example shows that the diaphragm flexibility has little influence on how the shears are distributed. However, this behavior is not necessarily applicable to all buildings.

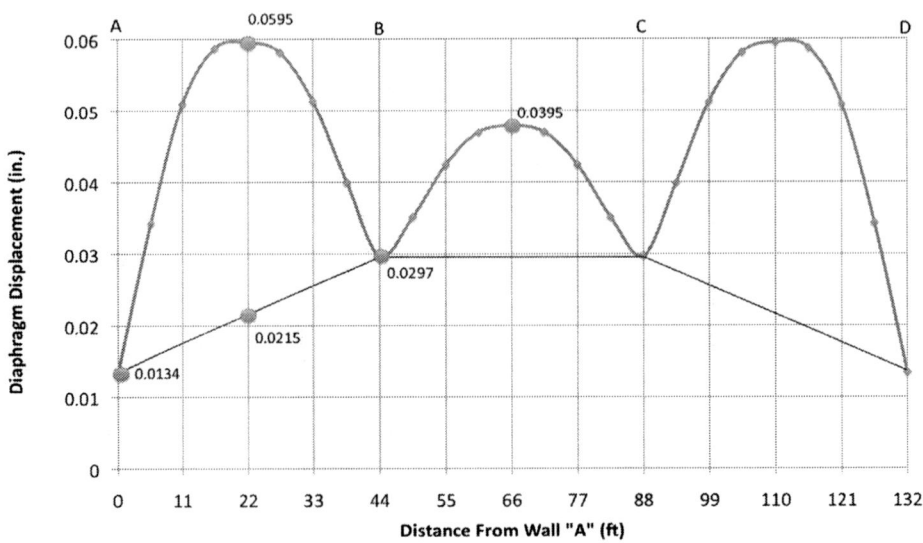

Figure G11–5 Diaphragm Displacements for a 2.0-in.-Thick Slab.

11.1 Accidental Torsion in Systems with Semirigid Diaphragms

Section 12.7.3 of ASCE 7 states that structures with semirigid diaphragms must be modeled to include the representation of the diaphragm stiffness (flexibility) and that additional degrees of freedom, aside from the two lateral displacements and one rotation at each level, must be included in the model. Clearly, any model that includes these effects is three-dimensional, so the de facto requirement of ASCE 7 is that systems with semirigid diaphragms must be modeled in three dimensions and presumably using a finite element approach wherein the diaphragm is discretized into a number of shell elements, as shown in **Fig. G11–3**. The question now arises as to whether the equivalent lateral force (ELF) method of analysis may be used to analyze the structure. Table 12.6-1 is silent on this issue because the only factor that excludes the possibility of using ELF, in Seismic Design Categories D through F, is a fundamental period greater than 3.5 T_S, or a system with T less than 3.5 T_S that has a horizontal irregularity of Type 1a or 1b or that has a vertical irregularity of Types 1a, 1b, 2, or 3.

In the author's opinion, the use of ELF for structures with semirigid diaphragms is not always appropriate, particularly if the diaphragm is somewhat flexible or of a highly irregular shape. In ELF analysis, the lateral loads are applied at the center of mass, and if the diaphragms are modeled using shell elements, there are considerable local deformations and stress concentrations at the location of the applied load. The deformation and stress patterns in the actual diaphragm are quite different because the inertial forces in the diaphragm are distributed throughout the diaphragm, not at a single point. Some improvement would be obtained if the ELF story forces were applied to the diaphragms in a distributed manner. However, it is not clear what this pattern should be because the pattern depends on the total accelerations at each point in the diaphragm, and this pattern is unknown at the beginning of the analysis. A modal response spectrum analysis or a modal response history analysis produces realistic distributions of inertial forces (if the diaphragm masses are distributed throughout the diaphragm and if a sufficient number of modes are used in the analysis) and is therefore more appropriate than ELF when the diaphragms are semirigid.

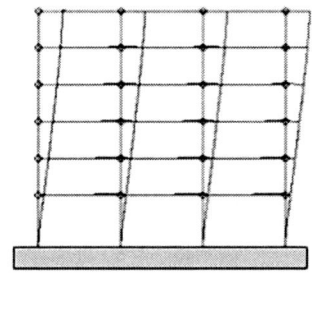

Example 12
Structural Analysis Requirements

This example covers the selection of the structural analysis procedure (Section 12.6) and requirements for modeling the structure (Section 12.7).

12.1 Selection of Structural Analysis Procedure

Section 12.6 and Table 12.6-1 provide the requirements for selection of the structural analysis procedure to be used to determine displacements, drifts, and member forces caused by seismic load effects. Three basic procedures are provided in Table 12.6-1:

1. equivalent lateral force (ELF) analysis
2. modal response spectrum (MRS) analysis
3. linear or nonlinear response history (LRH or NRH) analysis

Under certain circumstances, the simplified procedure provided in Section 12.14 may be used.

Table 12.6-1 shows that MRS and LRH (or NRH) analysis may be used for any system, and the ELF method may be used for almost any system. In fact, the essence of Table 12.6-1 can be restated as follows: The equivalent lateral force method of analysis may be used for any system, with the following exceptions:

1. structures in Seismic Design Categories D, E, or F with $T < 3.5\ T_S$ and with any of the following irregularities:

   Type 1 horizontal irregularity (torsional or extreme torsional)

   Type 1 vertical irregularity (soft story and extreme soft story)

   Type 2 vertical irregularity (weight or mass)

   Type 3 vertical irregularity (vertical geometric)

2. structures in Seismic Design Categories D, E, or F with $T \geq 3.5 T_S$

In terms of pure practicality, the ELF method of analysis should be used whenever it is permitted. Although the MRS method generally produces more accurate results than ELF, these results come at the expense of losing the signs (positive or negative) of seismic displacements and member forces, thereby complicating the combination of lateral seismic and gravity effects. The use of LRH analysis eliminates the problem with signs but requires much more effort on the part of the analyst, particularly as related to selection and scaling of ground motions. (See Example 6 in this guide for a description and discussion of the scaling process.) The complexity of NRH analysis is such that it should not be used except for special or important structures.

A potential advantage of MRS over ELF occurs when the computed system period of vibration is significantly greater than the upper limit period $C_u T_a$. This upper limit, described in Section 12.8.2, is used to provide a lower bound on the design base shear that is used in ELF, even when the computed period is greater than $C_u T_a$. When MRS analysis is used, Section 12.9.4 requires that the results be scaled such that the design base shear is not less than 85 percent of the base shear determined using ELF and the upper limit period. It is possible, therefore, that the system analyzed with MRS could be designed for 15 percent less base shear than the same system designed using ELF.

It is also important to note that some form of ELF analysis is required in the analysis and design of all building structures. For example, ELF is almost certainly used in preliminary design, where checks for torsional irregularities and soft story irregularities require the development of static forces along the height of the building. Additionally, diaphragm forces (Section 12.10 and Equation 12.10-1) are based on ELF story forces.

12.2 Examples for Computing T_S and 3.5 T_S

The purpose of this example is to determine those circumstances under which it is likely that $T > 3.5\ T_S$, thereby disallowing the use of ELF for structures that have been assigned to Seismic Design Categories D, E, and F. We base the example on a site with mapped MCE spectral accelerations $S_s = 0.8\ g$ and $S_1 = 0.25\ g$. Soil site classes B, C, D, and E are also considered. The quantities T_S and $3.5\ T_S$ are determined, as well as the height and approximate number of stories of steel moment frame and braced frame buildings that would have a period of $3.5\ T_S$. Heights are based on the following:

$$T = 3.5\ T_S = C_u T_a \quad \text{(Section 12.8.2)}$$

where $C_u = 1.4$ from Table 12.8-1

$$T_a = C_t h_n^x \quad (12.8\text{-}7)$$

According to Table 12.8-2, C_t and x are, respectively, 0.028 and 0.8 for steel moment-resisting frames, and 0.02 and 0.75, respectively, for concentrically braced frames.

T_S is defined in Section 11.4.5 and is computed as follows:

$$T_S = \frac{S_{D1}}{S_{DS}}$$

Physically, T_S is the period at which the constant acceleration (Eq. 12.8-3) and the constant velocity (Eq. 12.8-4) branches of the response spectrum meet (Fig. 11.4-1).

The results of the calculations are presented in **Table G12–1**. The T_S values are given in column 6, and $3.5\ T_S$ values in column 7. Clearly, the $3.5\ T_S$ values increase with increasing softness of the soil. This phenomenon occurs because the site amplification factor F_v is always greater than or equal to the factor F_a.

Columns 8 and 9 of **Table G12–1** show the heights of the structures in feet that give periods equal to $3.5\ T_S$. Also shown in parentheses is the number of stories, assuming that all stories have a height of 12.5 ft. The values in column 9, for the braced frame, are particularly interesting because the heights for all site classes are greater than the height limit of 160 ft given in Table 12.2-1 for a special concentrically braced frame. The conclusion that may be drawn from this is that the ELF method of analysis may be used for all regular concentrically braced frames (except for a number of structures in site class B, taller than 207 ft, which have height limits extended to 240 ft as allowed by Section 12.2.5.4). Regarding regular special moment frames, the values in column 8 of **Table G12–1** indicate that the ELF method may be used for any regular structure that is less than approximately eight stories high.

Table G12–1 T_s Values for SDC D Structures on Various Sites

1	2	3	4	5	6	7	8	9
Site	F_a	F_v	S_{DS} (g)	S_{D1} (g)	T_s (s)	$3.5\,T_s$ (s)	h_n M.F. (ft) (No. of Stories)	h_n B.F. (ft) (No. of Stories)
B	1.00	1.00	0.533	0.167	0.313	1.094	97 (8)	207 (16)
C	1.08	1.55	0.576	0.258	0.448	1.570	153 (12)	336 (26)
D	1.18	1.90	0.629	0.317	0.503	1.76	177 (14)	359 (28)
E	1.14	3.00	0.608	0.500	0.822	2.878	327 (26)	754 (60)

Note: Table based on $S_s = 0.8\,g$ and $S_1 = 0.25\,g$. M.F. = moment frame; B.F. = braced frame.

12.3 Structural Analysis Considerations

Aside from the method of analysis, the other principal modeling decisions that need to be made are the modeling of roof and floor diaphragms and whether three-dimensional analysis is required.

Section 12.7.3 requires that 3D modeling be used when horizontal structural irregularities of Types 1, 4, or 5 of Table 12.3-1 exist. When a 3D model is required, the diaphragms may be modeled as rigid only if the diaphragms are rigid in accordance with Section 12.3.1.2. Otherwise, the diaphragms must be modeled as semirigid.

Section 12.7.3 provides a host of other modeling requirements, such as inclusion of P-delta effects, representation of strength and stiffness of nonstructural components, inclusion of cracking in concrete and masonry structures, and modeling of deformations in the panel zone region of steel moment frames. A brief discussion of these issues is provided in Example 19. However, considerable thought and experience are required for the development of analytical models for structural analysis. The more advanced the analysis, the more detail required in the model and the more time and thought required to develop a reasonable mathematical model. Additionally, validation of complex models is much more difficult than it is for simpler models.

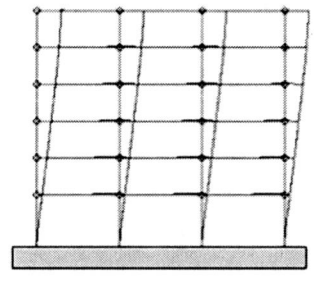

Example 13
Determining the Redundancy Factor

Section 12.3.4 describes the methodology for determination of the redundancy factor, ρ, which is used in several of the seismic load combinations that are specified in Section 12.4. This example demonstrates how the redundancy factor is determined for a variety of structural systems. The use of the redundancy factor in the context of the load combinations is demonstrated in Example 18 of this guide.

The value of the redundancy factor is either 1.0 or 1.3, depending on the Seismic Design Category (SDC) and the structural configuration. Section 12.3.4 indicates that the redundancy factor may be different in the two orthogonal loading directions of the structure.

For structures in SDCs B and C, ρ is taken as 1.0 in each direction. Section 12.3.4.1 provides a list of other situations in which the redundancy factor may be taken as 1.0. These situations (e.g., ρ = 1.0 in drift calculations) are applicable for all SDCs. Note, however, that Section 12.12.1.1 requires that drift limits be multiplied by $1 \neq \rho$ for moment frames in SDCs D, E, and F.

For structures in SDCs D, E, and F, the redundancy factor is 1.3 unless it can be shown that certain conditions exist, as outlined in Section 12.3.4.2 and Table 12.3-3. Condition (b) in Section 12.3.4.2 provides a description of situations in which the redundancy factor may be taken as 1.0 in both directions. If these conditions are not met, then condition (a) must be evaluated, from which the redundancy factor is either 1.0 or 1.3 in the direction of interest.

Condition (b) states that ρ may be taken as 1.0 where the structure has no plan irregularities and where there are two bays of perimeter seismic force resisting elements on each side of the building for each story of the building resisting more than 35 percent of the base shear. One or more collinear shear walls may be considered a bay only if the total plan length of the wall (or walls) is greater than the story height.

Fig. G13–1 illustrates several cases for which condition (b) may be evaluated. The floor plans in the figure are applicable at levels for which the design shear is greater than 35 percent of the base shear. Only building A in the figure satisfies the condition (b) test. The walls on each side of this plan-regular building are long enough to be classified as a bay, and there are two perimeter bays on each side of the building. Building B violates the criteria because the two walls marked with asterisks are not on the perimeter. Additionally, the system has a nonparallel system irregularity because the walls are not symmetrically placed about the center of resistance. Building C, which is assumed to have no irregularities, does not satisfy the criteria in the Y direction because the plan length of the four walls marked with asterisks is insufficient to classify the walls as bays. Buildings D, E, and F cannot automatically be classified with $\rho = 1.0$ because each has a plan irregularity. In building F, the plan irregularity occurs because of an out-of-plane offset of the shear walls marked with asterisks. The walls at the upper level are on the interior of the building and transfer to the exterior at the lower levels.

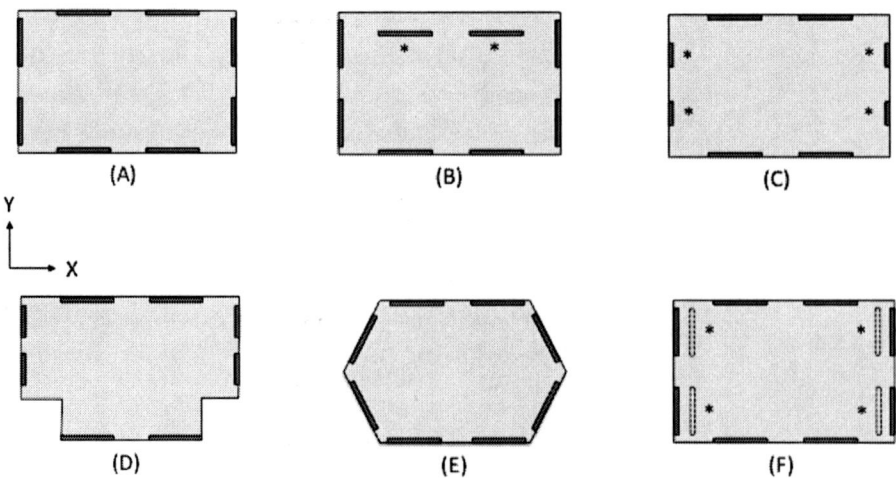

Figure G13–1 Evaluation of the Redundancy Factor for Various Buildings.

The fact that condition (b) has not been satisfied does not mean that the redundancy factor is 1.3. This situation would be the case only if condition (a) in Section 12.3.4.2 is also not met. Consider again building B of **Fig. G13–1**. In this shear wall system, each of the walls has a plan length greater than the height of the wall. Thus, the height-to-width ratio of the walls is less than 1.0, and the system defaults to "Other" lateral force resisting elements in Table 12.3-3. Presumably, therefore, this system can be assigned a redundancy factor of 1.0 in each direction because no requirements dictate otherwise. It appears that the same situation would occur even if the walls marked by asterisks in building B of **Fig. G13–1** were removed entirely. In the opinion of the author, this situation violates the spirit of the redundancy factor concept, and a factor of 1.3 should be assigned in this case.

In building C of **Fig. G13–1**, each of the walls marked with an asterisk has a length less than the story height. Removal of one of these walls does not cause an extreme torsional irregularity. It would appear at first glance that the removal of one wall would reduce the strength of the system by only 25 percent in the Y direction. However, this situation does not consider the effect of torsion. The reduction in strength must be based on the questions "How much lateral load can be applied in the Y direction for the system with one wall missing, and how does that compare to the strength of the system with the wall in place?"

Two interpretations exist for evaluation of the strength of the system with elements removed. The first is based on elastic analysis, and the second is based on inelastic analysis. An important consideration of the use of an inelastic analysis is that the system must be sufficiently ductile to handle the continued application of loads after the lateral load resisting elements begin to yield.

Consider, for example, the system shown in **Fig. G13–2**. This system has eight identical walls, marked A through H, each with a force–deformation relationship as shown in **Fig. G13–3**. The lateral load carrying capacity of each wall is 100 kips.

The system is evaluated on the basis of the following situations:

1. elastic behavior with all walls in place
2. elastic behavior with one wall removed
3. inelastic behavior analysis with all walls in place
4. inelastic behavior with one wall removed

For each situation, the lateral force V is applied in the Y direction at an eccentricity of 5 percent of the width (eccentricity = 0.05 × 125 ft = 6.25 ft) of the building in the X direction. The eccentricity of 5 percent of the plan width is consistent with the accidental torsion requirements of Section 12.8.4.2.

The analysis was performed for this example by use of a computer program that can model inelastic structures. This analysis provided the curves

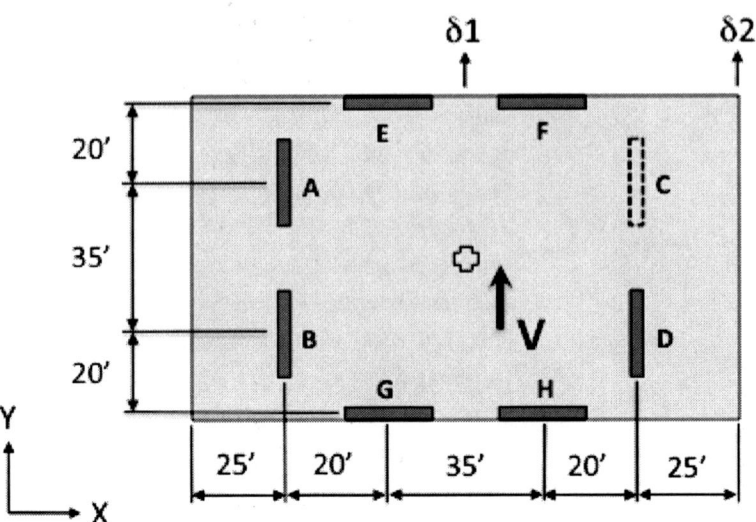

Figure G13–2 System with One Wall (C) Removed.

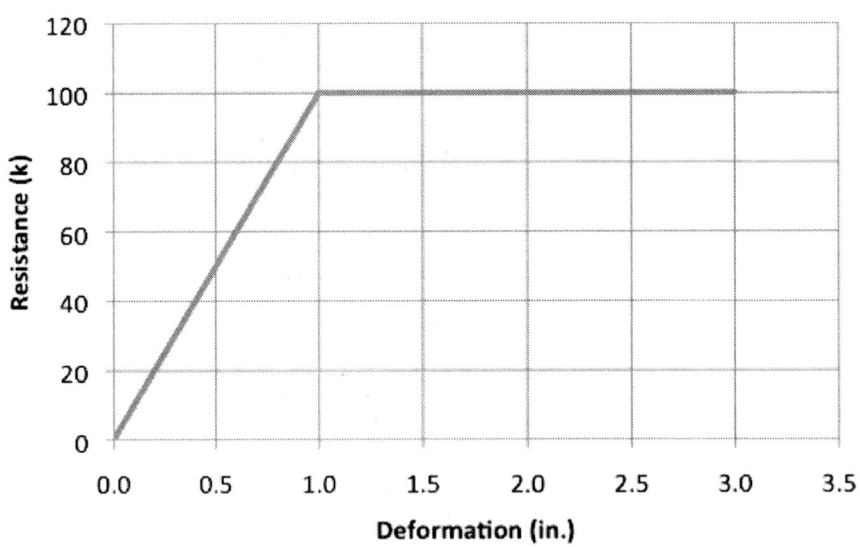

Figure G13–3 Force–Deformation Relationship for Shear Wall.

shown in **Fig. G13–4**. The upper curve represents the behavior of the system with all four walls in place, and the lower curve is for the system with wall C removed. The elastic analysis for each system is represented by the response up to first yield (the first change in slope of the curves), and the inelastic response is represented by the full curve.

From the perspective of the elastic analysis, the structure with all walls in place can resist a lateral load V of 370 kips. At this load, walls C and D on the right side of the building carry 100 kips, and walls A and B carry 85 kips.

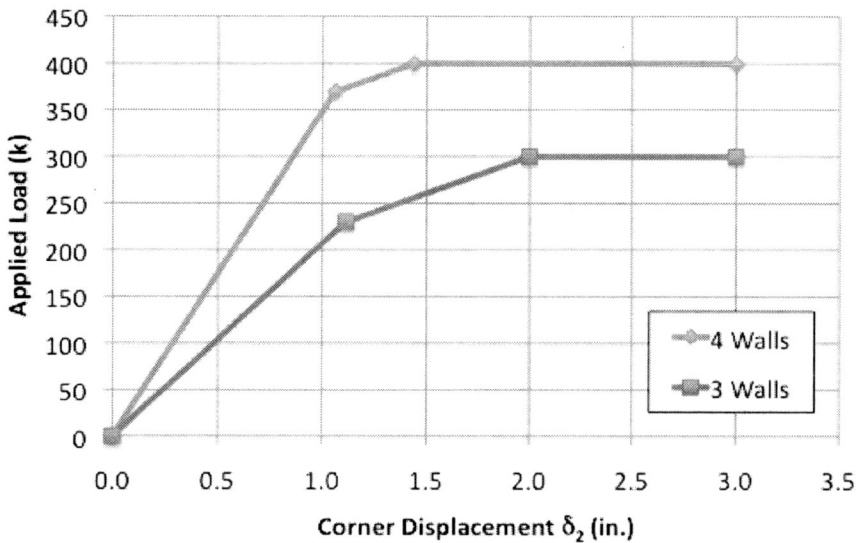

Figure G13–4 Force–Deformation Plot for Structure with Three or Four Walls Using Inelastic Analysis.

The ratio of the displacement $\delta 2$ relative to the displacement at point $\delta 1$ is 1.14, so according to Table 12.3-1, the structure does not have a torsional irregularity.

When the inelastic response is considered, the structure can carry additional lateral load because walls A and B can each resist an additional 15 kips before they reach their 100-kips capacity. With four walls resisting 100 kips each, the total lateral capacity of the system is 400 kips. Recall that the force–displacement plot for this four-wall system is shown by the upper curve in **Fig. G13–4**. The displacement shown in the figure is the Y direction displacement $\delta 2$. The first change of slope in the curve occurs when walls C and D yield, and the second change occurs when the walls A and B yield.

When wall C is removed, the center of rigidity moves 9.375 ft to the left. When an elastic analysis is performed, the system can resist a lateral load of only 230 kips because at this point wall D reaches its 100-kips capacity. At this point, walls A and B resist 65 kips each. Additionally, the ratio of the displacement at point $\delta 2$ with respect to the displacement at point $\delta 1$ is 1.35. Hence, a torsional irregularity, but not an extreme irregularity, exists. The ratio of the resisting force of the three-wall system to that of the four-wall system is 230/370 or 0.621. On this basis, the system must be designed with $\rho = 1.3$ because it loses more than 33 percent of its strength.

From the perspective of an inelastic analysis, the three-wall system can resist a total of 300 kips. This is shown by the force–displacement diagram (the lower curve) in **Fig. G13–4**. The ratio of the inelastic resisting force of the three-wall system to that of the four-wall system is $300/400 = 0.75$. In this case, the system could theoretically be designed with $\rho = 1.0$ because it passes both the strength and the nonextreme torsion irregularity tests.

The inelastic strength ratios for the system with and without one wall removed could be obtained a priori, by simply summing the wall strengths in the direction of loading. Thus, it is not necessary to obtain the full inelastic force–deformation responses shown in **Fig. G13–4**. It is necessary, however, to perform sufficient analysis to determine whether the removal of one wall causes an extreme torsional irregularity.

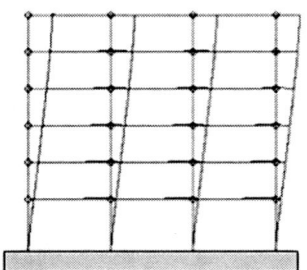

Example 14

Accidental Torsion and Amplification of Accidental Torsion

A variety of issues related to torsional loading are considered in this example. Included are torsional irregularities, accidental torsion, torsional amplification, and application of accidental torsion to structures analyzed using the equivalent lateral force or modal response spectrum approach. Emphasis is on structures with rigid diaphragms; however, some guidance is provided for torsional analysis of structures with semirigid diaphragms. Accidental torsion need not be considered for systems with flexible diaphragms.

Fig. G14–1 is a plan view of a five-story reinforced concrete shear wall building. The first story is 12 ft, 4 in. tall, and the upper stories are each 11 ft, 4 in. tall. The building, located in central Missouri, is used to house a variety of business offices and is classified as Occupancy Category II. The following design spectral accelerations have been determined for the site:

$S_{DS} = 0.45\ g$
$S_{D1} = 0.19\ g$

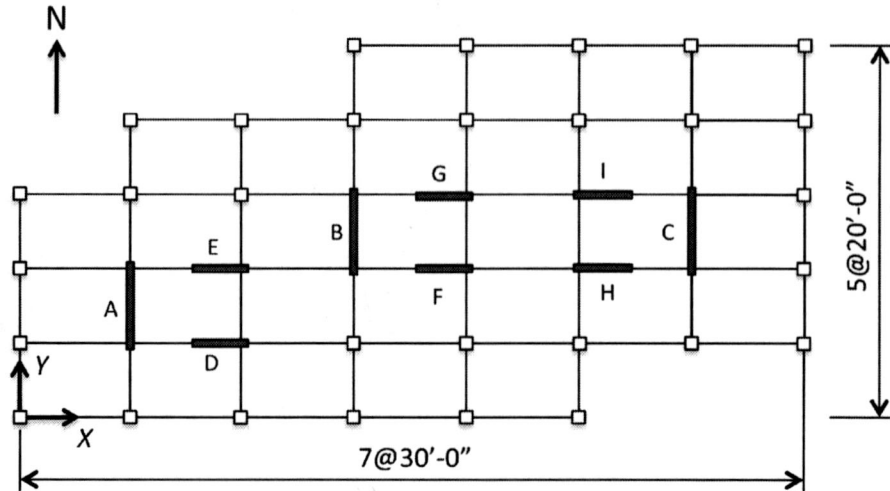

Figure G14–1 Plan of Reinforced Concrete Shear Wall Building.

Tables 11.6-1 and 11.6-2 indicate that the Seismic Design Category is C. Table 12.2-1 allows the use of an ordinary reinforced concrete shear wall, with the following design parameters:

$R = 5$
$C_d = 4.5$

For this analysis, it is assumed that all walls are 10 in. thick and are constructed with 4,000 lb/in.² normal-weight concrete. Walls A, B, and C have a length of 22 ft, and walls D, E, F, G, H, and I have a length of 16 ft.

The period of vibration is estimated from Eq. 12.8-7. Using h_n = 12.33 + 4(11.33) = 57.7 ft and coefficients C_t = 0.02 and x = 0.75 from Table 12.8-2,

$$T_a = C_t h_n^x = 0.02 \times 57.7^{0.75} = 0.42 \text{ s}$$

This period is used for analysis because an analytical period (from a computer program) is not available.

Lateral forces are computed using the equivalent force method. Using the ground motion parameters given above, $T_S = S_{D1}/S_{DS} = 0.19/0.45 = 0.42$, which by coincidence is equal to $T_a = 0.42$ s, so Eqs. 12.8-2 and 12.8-3 produce the same base shear. Using a total seismic weight of the building of 9,225 kips, the base shear is determined from Eq. 12.8-2:

$$C_s = \frac{S_{DS}}{\left(\frac{R}{I}\right)} = \frac{0.45}{\left(\frac{5.0}{1.0}\right)} = 0.090$$

$$V = C_s W = 0.090(9{,}225) = 830 \text{ kips}$$

Equivalent lateral forces are computed in accordance with Section 12.8.3, with $k = 1.0$. The results of the ELF calculation are provided in **Table G14–1**.

According to Section 12.3.1.2, the structure may be analyzed with a rigid diaphragm assumption. For such structures, by requirement of Sections 12.8.1 and 12.8.2, inherent torsion and accidental torsion must be included in the analysis. There is no need to compute separately the effects of inherent torsion because this computation is automatically included in a three-dimensional analysis, which is required for this structure because the structure has, at the minimum, a nonparallel system irregularity. This irregularity occurs because the lateral load resisting elements (walls) are not symmetric about the major orthogonal axes of the lateral load resisting system (Table 12.3-1). It is possible that the structure also has a torsional irregularity. If a torsional irregularity exists, it is likely that the accidental torsion needs to be magnified for this SDC C building, in accordance with Section 12.8.4.3.

A 3D analysis is used to determine if there is a torsional irregularity and whether the accidental torsion must be magnified. The analysis was run using a 3D finite element analysis program, wherein the walls were modeled as membrane elements. Membrane elements were used in lieu of shell elements because it was desired for simplicity to exclude the out-of-plane stiffness of the walls. The use of membrane elements automatically includes in-plane axial, flexural, and shear deformation in the walls. Uncracked properties were used because the main purpose of the analysis is to determine the elastic deformations in the system and to determine the distribution of the forces in the walls. Although the absolute magnitude of displacements is affected by cracking, the ratio of the displacements at the edge of the building to the displacement at the center of the building is not affected, as long as the same stiffness reduction factors are used to represent cracking in each wall. Similarly, the distribution of forces is not affected by cracking if all walls are cracked to the same degree.

Table G14–1 Equivalent Lateral Forces

1	2	3	4	5	6	7
Level	H (ft)	h (ft)	W (kips)	Wh^k	Wh^k/Total	F (kips)
5	11.33	57.66	1,820	104,941	0.326	271
4	11.33	46.33	1,845	85,479	0.266	221
3	11.33	35.00	1,845	64,575	0.201	167
2	11.33	23.66	1,845	43,653	0.136	113
1	12.33	12.33	1,870	23,057	0.070	58
Total	57.65	—	9,225	321,705	1.00	830

(This discussion is based on flexural properties and flexural cracking. For shear wall systems, the effect of shear deformations and shear cracking should also be considered because these effects can have a significant influence on the distribution of forces in the system.)

For this analysis, the diaphragms were modeled with artificially thick membrane elements to make the results consistent with the rigid diaphragm assumption. Because of this approach, the diaphragms were not in fact rigid, but almost rigid. Fully rigid diaphragms can be obtained through the use of mathematical constraints. The topic of diaphragm rigidity is discussed in more depth later in this example.

The analysis is carried out only for forces acting in the north–south direction. Three load conditions are applied, one without accidental eccentricity, one with the lateral force applied east of the center of mass, and the other with the forces applied west of the center of mass. The location of the center of mass is shown in **Fig. G14–2**.

As required by Section 12.8.4.2, the lateral forces are applied at an eccentricity of 0.05 times the length of the building perpendicular to the direction of loads. Thus the eccentricity is $0.05(210) = 10.5$ ft when the lateral loads are applied in the north–south direction. Because of the rigid diaphragm assumption, the lateral load, including torsion, may be applied through the use of two loading points, shown as F1 and F2 in **Fig. G14–2**. (The loads could also be applied as a single concentrated force (lateral load) plus moment applied about the vertical axis (accidental torsion).

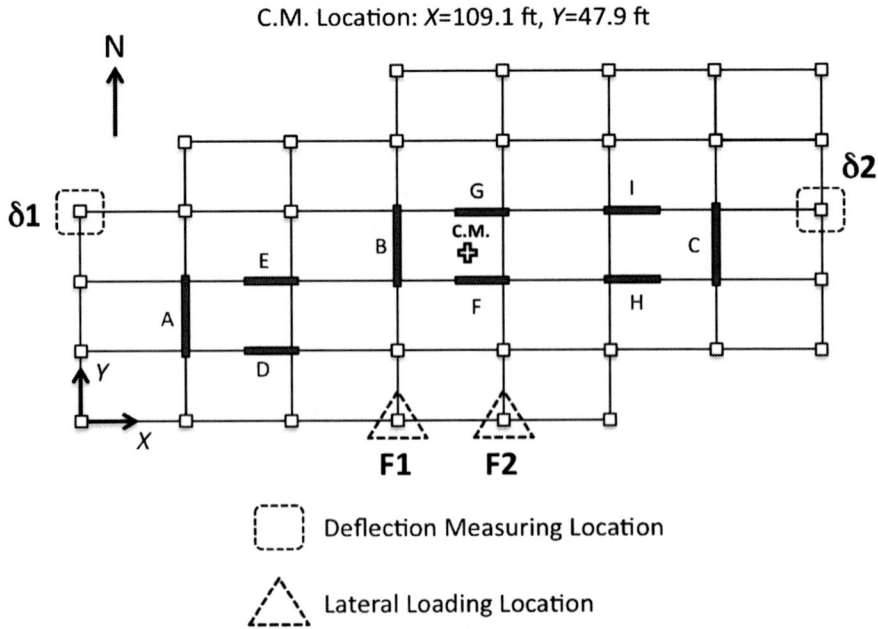

Figure G14–2 Plan Shown with Loading and Monitoring Locations. C.M. is center of mass.

These loads would be applied at a single node at the center of mass.) The forces caused by the lateral load without torsion are based on simple beam reactions for a beam with a span of 30 ft, as shown in **Fig. G14–3(a)**. The accidental torsional component of load is also applied as two concentrated forces, shown in **Fig. G14–3(b)**. The total load is simply the lateral load, plus or minus the torsional load, as illustrated in **Figs. G14–3(c)** and **(d)**.

To determine if a torsional irregularity exists (Table 12.3-1), interstory drifts (not displacements) are monitored at the extreme edges of the building under a loading that consists of the design lateral forces and (plus or minus) accidental torsion. If the maximum drift at the edge of any story exceeds 1.2 times the average of the drifts at the two edges of the story, a torsional irregularity exists. If the maximum drift exceeds 1.4 times the average, an extreme irregularity exists. In this example, the computed displacements were used without the deflection amplifier C_d because the C_d term cancels out when computing the ratio of drifts.

For determining the torsional amplification factor, the story displacements (not drifts) are used. An amplification factor, A_x, is then computed for each story. This factor is determined using Eq. 12.8-14:

$$A_x = \left(\frac{\delta_{max}}{1.2\delta_{avg}}\right)^2$$

where δ_{max} is the maximum deflection at the edge of the story, and δ_{avg} is the average deflection at the two edges. The two points at which the deflections were monitored are designated as $\delta 1$ and $\delta 2$ in **Fig. G14–2**.

Inherent torsion must be included in the analysis, which is automatically accomplished when a 3D analysis is performed. There is no need to separate out the inherent torsion, however, because it is never used independently. Nevertheless, it is worthwhile to apply the lateral loads without accidental torsion and observe the resulting deflection patterns. These deflections give an indication of the location of the center of rigidity relative to the center of mass. Accidental torsion loadings that rotate the floor plates in the same direction as the rotation resulting from inherent torsion clearly control when determining if torsional irregularities occur and when computing amplification factors.

To determine the displacements in the system under lateral load only, the forces listed in column 7 of **Table G14–1** were applied as shown in **Fig. G14–3(a)**. The results of the analysis are presented in **Table G14–2** The rotational measurement, θ, is the rotation about the vertical axis, which is counterclockwise positive. The deflections at point $\delta 2$ at any level are always greater than those at $\delta 1$, causing the building to twist counterclockwise, which is a positive rotation. This situation indicates that the center of mass lies east of the center of rigidity, which is to the right in **Fig. G14–2**. Clearly, lateral loads plus torsion results in greater twisting in the positive direction.

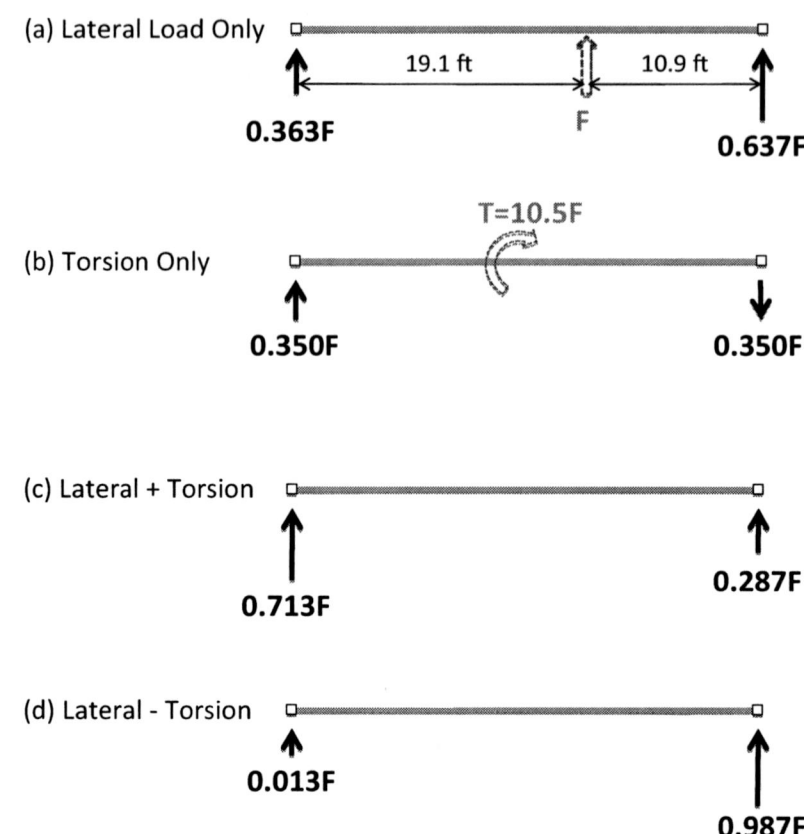

Figure G14–3 Loading Values Applied to Building.

Table G14–2 Displacements for Building Under Lateral Force Without Torsion

1	2	3	4
Level	δ_1 (in.)	δ_2 (in.)	θ (radians)
5	0.463	0.754	1.15E – 04
4	0.341	0.555	8.50E – 05
3	0.224	0.364	5.56E – 05
2	0.120	0.196	2.98E – 05
1	0.042	0.068	1.04E – 05

The results in **Table G14–3(a)** indicate an increasing rotation when the torsion is applied in the positive direction. The twisting causes an extreme torsional irregularity. A given irregularity type needs to occur only at one story for the whole building to be classified as having that irregularity. **Table G14–3(b)** shows the computation of the torsional amplification factor at each level. In this case, the values are virtually the same all the way up the building because the walls continue the full height of the building.

Table G14–3a Results for Lateral Plus Accidental Torsion: Irregularity Check

1	2	3	4	5	6	7	8
Level (δ) Story (Δ)	δ_1 (in.)	δ_2 (in.)	Δ_1 (in.)	Δ_2 (in.)	Δ_{avg} (in.)	$\Delta_{max}/\Delta_{avg}$	Torsional Irregularity
5	0.304	0.929	0.080	0.245	0.163	1.508	Extreme
4	0.224	0.684	0.077	0.235	0.156	1.510	Extreme
3	0.147	0.449	0.068	0.208	0.138	1.505	Extreme
2	0.079	0.241	0.052	0.157	0.104	1.507	Extreme
1	0.027	0.084	0.027	0.084	0.056	1.506	Extreme

Table G14–3b Results for Lateral Plus Accidental Torsion: Amplification Factors

1	2	3	4	5	6
Level (δ)	δ_1 (in.)	δ_2 (in.)	δ_{avg} (in.)	$\delta_{max}/\delta_{avg}$	A_x
5	0.304	0.929	0.617	1.508	1.578
4	0.224	0.684	0.454	1.507	1.578
3	0.147	0.499	0.298	1.507	1.577
2	0.079	0.241	0.160	1.507	1.577
1	0.027	0.084	0.056	1.506	1.576

When the torsion is applied in the opposite direction, it offsets the inherent torsion, and the irregularity disappears. Additionally, the computed amplification factors are less than 1.0, so the minimum factor of 1.0 is applied. If the torsional irregularity occurs for any direction of the applied eccentricity, the entire system is classified as having a torsional irregularity. These results are provided in **Table G14–4**.

A few points are worthy of discussion at this point:

1. When determining torsional amplification, it is not necessary to iterate by analyzing the system with the amplified accidental torsion, determining a new amplification factor, analyzing again, and so on. Amplification factors are determined by a single analysis using the 5 percent accidental eccentricity.
2. When assessing torsional regularity or torsional amplification, it is not necessary to apply the accidental torsion simultaneously in the two orthogonal directions.

Table G14–4a Results for Lateral Minus Accidental Torsion: Irregularity Check

1	2	3	4	5	6	7	8
Level (δ) Story (Δ)	$\delta 1$ (in.)	$\delta 2$ (in.)	$\Delta 1$ (in.)	$\Delta 2$ (in.)	Δ_{avg} (in.)	$\Delta_{max}/\Delta_{avg}$	Torsional Irregularity
5	0.623	0.579	0.165	0.153	0.159	1.038	None
4	0.458	0.426	0.157	0.146	0.152	1.036	None
3	0.301	0.280	0.139	0.130	0.135	1.033	None
2	0.162	0.150	0.106	0.098	0.102	1.039	None
1	0.056	0.052	0.056	0.052	0.054	1.037	None

Table G14–4b Results for Lateral Minus Accidental Torsion: Amplification Factors

1	2	3	4	5	6
Level (δ)	$\delta 1$ (in.)	$\delta 2$ (in.)	δ_{avg} (in.)	$\delta_{max}/\delta_{avg}$	A_x
5	0.623	0.579	0.601	1.037	1.0
4	0.458	0.426	0.442	1.036	1.0
3	0.301	0.280	0.291	1.036	1.0
2	0.162	0.150	0.156	1.038	1.0
1	0.056	0.052	0.054	1.037	1.0

Note: Minimum $A_x = 1.0$.

14.1 Applying Accidental Torsion in Systems with Semirigid Diaphragms

It was mentioned earlier in this example that the floor and roof diaphragms were modeled with artificially thick membrane elements. This modeling was done to represent a rigid diaphragm. It seems more logical, however, to model the diaphragms with their true thickness because this modeling provides a more accurate representation of the distribution of forces to the different lateral load resisting elements. Additionally, the analysis reports diaphragm stresses that can be used to determine collector and diaphragm connection forces. These issues are discussed in Example 21 of this guide.

If the true diaphragm thickness is used in the analysis, the lateral forces should not be applied as shown in **Fig. G14–2**. Instead, these forces should be distributed throughout the diaphragm in some reasonable manner. A possible loading for the structure analyzed in this example is shown in **Fig. G14–4**. In this figure, the lateral force at a level is applied on the basis of nodal forces, where the sum of the individual nodal forces is equal to the total load applied at the level. Nodal forces are based on a tributary area (mass) basis. Accidental torsion is applied by a series of moments applied in the plane of the diaphragm. Shell, not membrane, elements would be required for such an analysis, and the element formulation used in the finite element analysis program must be able to accommodate drilling degrees of freedom (nodal moments applied about an axis normal to the plane of the diaphragm). If it is not possible to apply moments directly to the nodes, the torsion may be applied by modifying the lateral forces that are applied to each node.

It is important to note, however, that indiscriminate use of ELF analysis with semirigid diaphragms is not appropriate. This is particularly true when the diaphragm tends to be more flexible than rigid or when the diaphragm has a highly irregular shape (e.g., the diaphragm in Example 9, **Fig. G9–6**).

14.2 Application of Accidental Torsion and Torsional Amplification in Modal Response Spectrum Analysis

The loading shown in **Fig. G14–4** would be useful in determining if torsional irregularities exist, computing torsional amplification factors, and for the application of the lateral loading in an equivalent lateral force analysis.

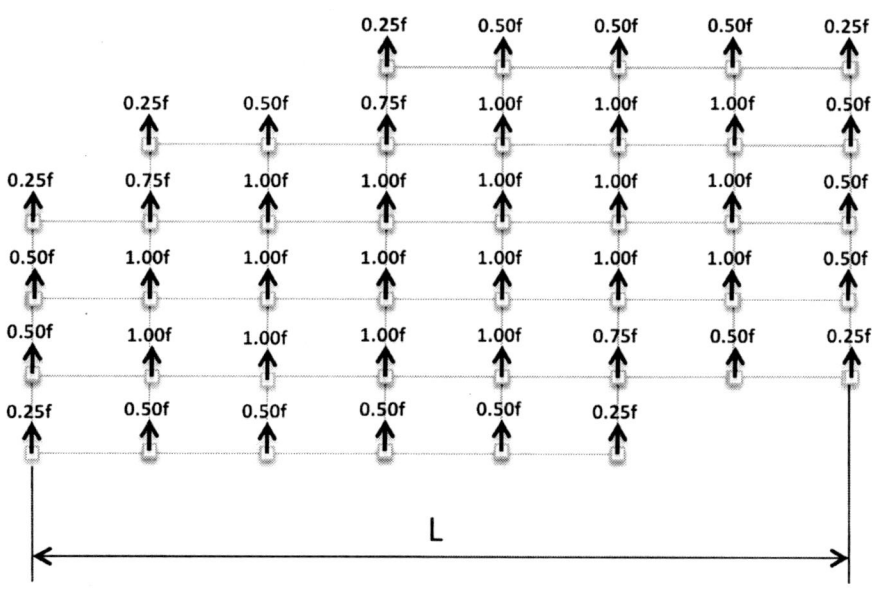

For accidental torsion apply a Z-direction moment of 0.05L times nodal force at each loading location.

Figure G14–4 Direct and Torsional Loading for a System with a Semirigid Diaphragm.

If a modal response spectrum (MRS) analysis is used, it is difficult to apply the accidental torsion in a truly dynamic manner (e.g., by physical adjustment of the center of mass). Doing so results in a different set of modal properties (frequencies and mode shapes) for each adjustment of mass. Instead, it is recommended that the effects of accidental torsion be included in a separate static analysis and then combined with the results of the MRS analysis. A similar approach is recommended when a rigid diaphragm assumption is applicable.

When applying the accidental torsion statically, equivalent lateral forces, scaled in accordance with Section 12.9.4, would be applied at a 5.0 percent eccentricity, exactly as is done in a full ELF analysis. Application of accidental torsion using an ELF force distribution is permitted even where an MRS analysis (or higher) is required by Table 12.6-1.

However, whenever static loading is used to apply the accidental torsion, it is necessary (for torsionally irregular buildings in SDC C and above) to amplify the accidental torsion according to the requirements of Section 12.8.4.3. This is required even when the main lateral load analysis is performed using MRS analysis because the accidental torsion effects are not included in the dynamic model per the requirements of Section 12.9.5. Accidental torsion would be included in the dynamic model only if the mass locations were physically adjusted.

Example 15
Load Combinations

This example explores the use of the strength design load combinations that include earthquake load effects. These load combinations, numbered 5 and 7 in Section 2.3.2 of ASCE 7, are then discussed in context with the requirements of Section 12.4. Also discussed in this example are requirements for including direction of loading (Section 12.5), accidental torsion (Section 12.8.4.2), and amplification of accidental torsion (Section 12.8.4.3).

Chapter 2 of ASCE 7 provides the required load combinations for both strength-based and allowable stress-based designs. This example covers only the use of the strength-based load combinations. There are seven basic load combinations provided in Section 2.3.2. Each member and connection of the structure must be designed for the maximum force or interaction of forces (e.g., axial force plus bending) produced by any one of these basic combinations. For any given member, such as a reinforced concrete girder, it might be found that different combinations control different aspects of the design. For example, load combination 2 in Section 2.3.2 might control the requirements for bottom reinforcement at midspan, whereas combination 5 controls requirements for top reinforcement at the ends of the member. A specific example of this circumstance is provided later. The remainder of this example concentrates on combinations 5 and 7, which are shown below:

Combination 5: $1.2D + 1.0E + 1.0L + 0.2S$
Combination 7: $0.9D + 1.0E + 1.6H$

In each of these combinations, the factor on earthquake load effects, E, is 1.0. This is due to the fact that the spectral design accelerations S_{DS} and S_{D1}, produced from the requirements of Chapter 11, are calibrated to be consistent with an ultimate load. Wind loads, on the other hand, are nominal, and when used in strength design, must be multiplied by a factor greater than 1.0 (in combinations 4 and 6 of Section 2.3.2) to elevate them to the ultimate load level. Also, the factor on live load in combination 5 may be reduced to 0.5 in most cases (See Exception 1 in Section 2.3.2). The adopted building code may provide load combinations that are different than those specified in ASCE 7 (e.g., the alternate basic combinations of ICC 2006). If so, these combinations must be used in lieu of the ASCE 7 requirements.

The snow load in combination 5 is always included when $S > 0$. There might also be a snow load effect in E in both combinations because the effective seismic weight, W, is required to include 20 percent of the design snow load when the flat roof snow load exceeds 30 lb/ft^2 (Section 12.7.2). See Exception 2 in Section 2.3.2 for comments related to the use of the pressure load H.

There are two ways in which the combinations are used. The first, covered in Section 12.4.2, is applicable to all elements and connections in the structure and may be considered the standard load combinations. The second, covered in Section 12.4.3, is for those special elements or connections that must be designed with the overstrength factor, Ω_o. Where required, the special load combinations supersede the standard combinations (item 6 in Section 12.3.4.1).

ASCE 7 provides several specific cases where the overstrength load combination must be used. For example:

- Section 12.2.5.2, which requires that the overturning resistance of a cantilever structure be designed with the overstrength factor.
- Section 12.3.3.3, which pertains to elements supporting discontinuous walls or frames.
- Section 12.10.2.1, which pertains to collector elements, their splices, and their connections to resisting elements.

Section 12.4.2 of ASCE 7 provides details on the standard seismic load effect. For use in load combination 5, the seismic load effect E is given as

$$E = E_h + E_v \tag{12.4-1}$$

and for use in load combination 7,

$$E = E_h - E_v \qquad (12.4\text{-}2)$$

where

$$E_h = \rho Q_E \qquad (12.4\text{-}3)$$

and

$$E_v = 0.2 S_{DS} D \qquad (12.4\text{-}4)$$

The term E_h represents the horizontal seismic load effect. The term ρ in Eq. 12.4-3 is the redundancy factor, computed in accordance with Section 12.3.4. This value is 1.0 for all buildings assigned to Seismic Design Category (SDC) B or C and is either 1.0 or 1.3 in SDC D through F. This factor applies to the entire structure but may be different in the two orthogonal directions. See Example 13 in this guide for details on determination of the redundancy factor.

The term E_v represents the effect of vertical ground acceleration, which is not considered explicitly elsewhere, with the exception of Section 12.4.4, which provides requirements for minimum upward forces in horizontal cantilevers for buildings in SDC D through F.

Q_E in Eq. 12.4-3 is the seismic effect on an individual member or connection. This value is produced by the seismic analysis of the structure and includes direct loading (e.g., application of equivalent lateral forces), accidental torsion and torsional amplification (if applicable), and orthogonal loading effects (if applicable). Q_E might represent, for example, a bending moment at a column support, an axial force in a bracing member, or a stress in a weld. In some cases, Q_E might represent an interaction effect, such as an axial-force bending moment combination in a beam column. In such cases, both the axial force and bending moment occur concurrently and should be taken from the same load combination.

When Eqs. 12.4-1 through 12.4-4 are substituted into the basic load combinations, the following detailed combinations for strength design are obtained:

Combination 5: $(1.2 + 0.2 S_{DS}) D + \rho Q_E + 1.0 L + 0.2 S$
Combination 7: $(0.9 - 0.2 S_{DS}) D + \rho Q_E + 1.6 H$

The use of these load combinations is illustrated in **Fig. G15–1** and **Fig. G15–2**. Each figure shows a simple frame with gravity and seismic loading. Snow and hydrostatic pressure loading are not present.

The top of **Fig. G15–1** shows only the gravity portion of the load, with the heavy gravity case shown at the upper left and the light gravity case shown at the upper right. At the bottom of the figure, the loading is shown for seismic effect acting to the east or to the west. Moment diagrams are

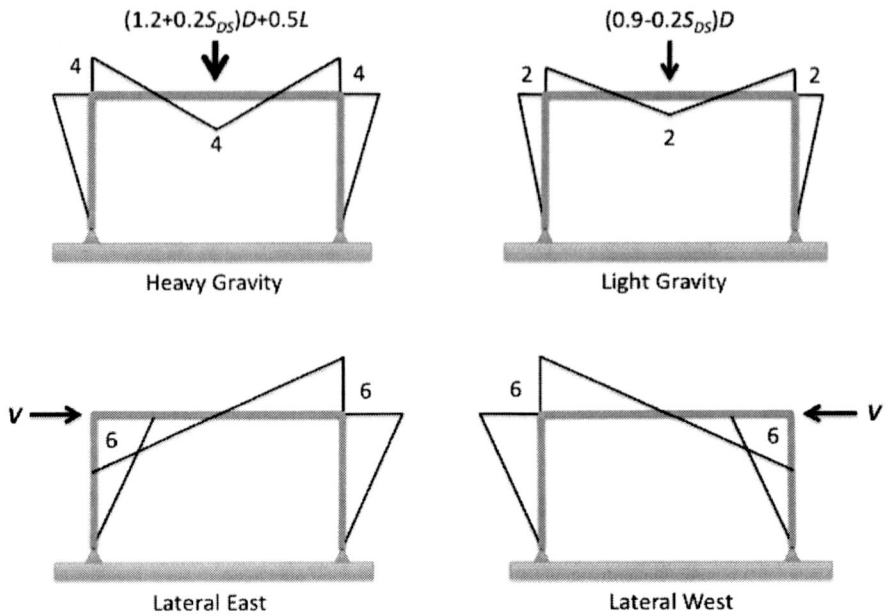

Figure G15–1 Basic Load Combinations for Simple Frame.

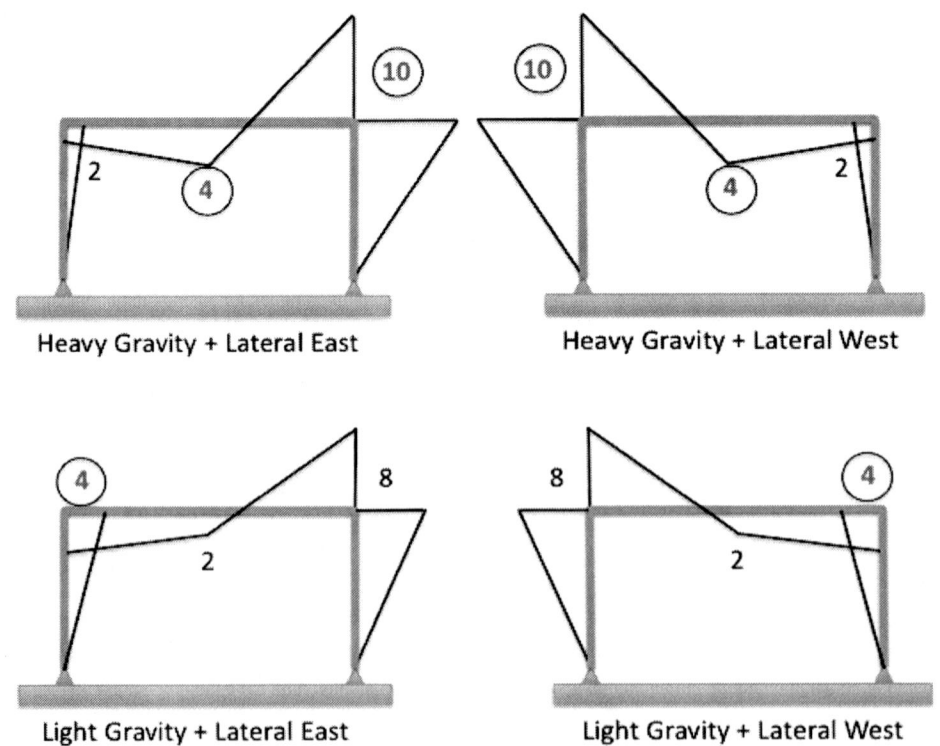

Figure G15–2 Combinations of Basic Combinations for Worst Effect.

drawn for each loading, and these diagrams are presented on the tension side. The moment values (units not important) are shown for each loading.

Fig. G15–2 shows the combination of gravity and earthquake load. The top of the figure gives the total moments for load combination 5, with heavy gravity plus seismic acting to the east on the left side of the figure and heavy gravity plus seismic acting to the west on the right side of the figure. The bottom of the figure is for combination 7 with light gravity and seismic acting to the east or west. Controlling moments are circled.

As may be seen in **Fig. G15–2**, each load combination must be exercised twice, once for positive seismic and once for negative seismic to produce the controlling effect in each member or connection. For the given example, the controlling tension on the top moment is 10 at both ends of the beam, and the controlling tension on the bottom moment is 4 for the full beam span. Of course, other load combinations (without seismic) provided in Chapter 2 of ASCE 7 must also be exercised to determine if they control. It is important to recognize, however, that the seismic detailing requirements associated with any system must be provided, regardless of the loading combination that controls the strength of the member or connection. For example, a member in an intermediate moment frame that has a wind-based design force twice as high as the seismic design force may be sized on the basis of the wind forces but must be detailed according to the requirements for intermediate moment frames.

The lateral forces shown in **Fig. G15–1** and **Fig. G15–2** include the effects of accidental torsion, as well as the effect of seismic loads acting simultaneously in orthogonal directions. Accidental torsion must be considered for any building with a nonflexible diaphragm, and torsionally irregular buildings in SDC C through F are subject to requirements for amplifying accidental torsion. Direction of load effects is covered in Section 12.5. For structures in SDC B, the analysis (including accidental torsion effects) may be performed independently in each direction and the structure may be designed on that basis. For structures in higher SDCs, the direction of load effects must be explicitly considered wherever a Type 5 horizontal nonparallel systems irregularity occurs (Table 12.3-1). Because such irregularities are common, many buildings must be designed for a complicated combination of loads that include gravity, lateral loads acting from any direction, simultaneous application of lateral loads acting in orthogonal directions, and accidental torsion with or without amplification.

The manner in which the different load effects are considered depends on whether the analysis is being performed using the equivalent lateral force (ELF) method or the modal response spectrum (MRS) method. Procedures used in association with response history analysis are beyond the scope of this guide.

Before illustrating the procedures used in association with ELF or MRS analysis, it is important to note that these procedures, as presented herein, depend on the analysis being performed in three dimensions. Section

12.7.3 requires 3D modeling for structures with horizontal structural irregularities of Type 1a, 1b, 4, or 5 of Table 12.3-1. Additionally, a 3D analysis is required for structures with semirigid diaphragms. Even where 3D analysis is not required by ASCE 7, it is advisable to use such analysis because the requirements for accidental torsion and loading direction are easier to apply than would be the case if the structure were to be decomposed into a number of 2D models.

15.1 Load Combination Procedures Used in ELF Analysis

In ELF analysis, as many as 16 seismic lateral load cases may be required. The generation of the 16 lateral load cases is shown in **Table G15–1** and in accompanying **Fig. G15–3**.

Table G15–1 Generation of ELF Load Cases

Major Load Direction	Major Load Applied at Eccentricity[a]	Orthogonal Load (applied at zero eccentricity)[b]	Load Case Number
$+V_{EW}$	$0.05 A_x B$	$+0.3\ V_{NS}$	1
		$-0.3\ V_{NS}$	2
	$-0.05 A_x B$	$+0.3\ V_{NS}$	3
		$-0.3\ V_{NS}$	4
$-V_{EW}$	$0.05 A_x B$	$+0.3\ V_{NS}$	5
		$-0.3\ V_{NS}$	6
	$-0.05 A_x B$	$+0.3\ V_{NS}$	7
		$-0.3\ V_{NS}$	8
$+V_{NS}$	$0.05 A_x L$	$+0.3\ V_{EW}$	9
		$-0.3\ V_{EW}$	10
	$-0.05 A_x L$	$+0.3\ V_{EW}$	11
		$-0.3\ V_{EW}$	12
$-V_{NS}$	$0.05 A_x L$	$+0.3\ V_{EW}$	13
		$-0.3\ V_{EW}$	14
	$-0.05 A_x L$	$+0.3\ V_{EW}$	15
		$-0.3\ V_{EW}$	16

a. A_x is the torsional amplification factor.
b. Not always required. See Section 12.5

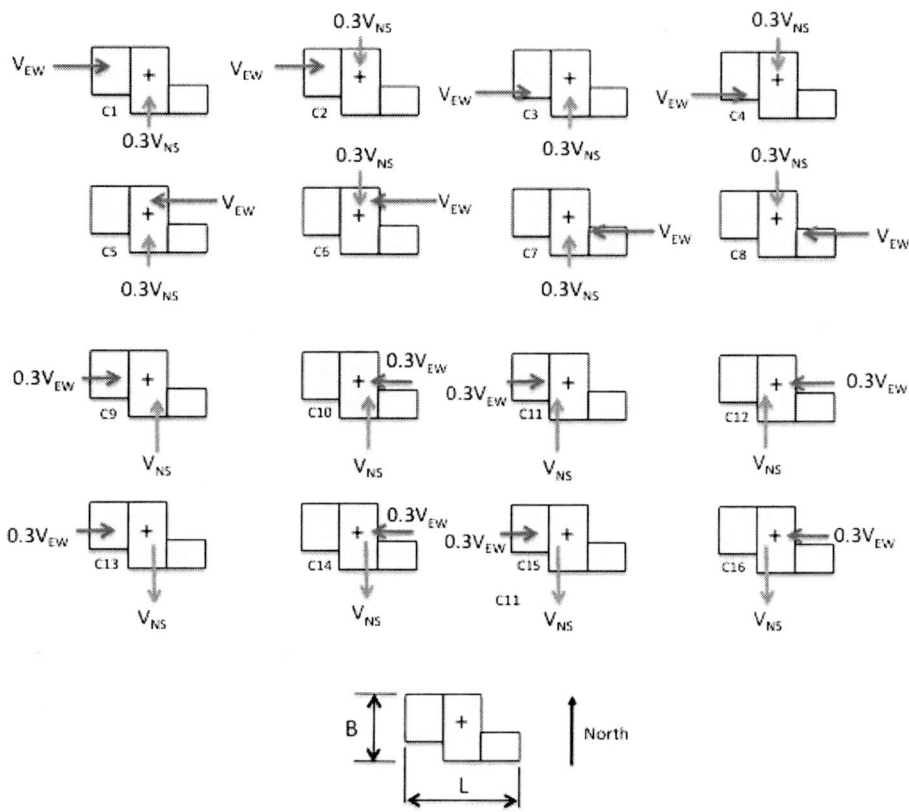

Figure G15–3 Sixteen Basic Lateral Load Cases Used in ELF Analysis.

The first column of **Table G15–1** represents the direct lateral load without accidental eccentricity. These forces would come from Eqs. 12.8-1, 12.8-11, and 12.8-12. The second column provides the eccentricity at which the lateral loads must be applied. As described in Section 12.8.4.2, this eccentricity is equal to at least 0.05 times the dimension of the building perpendicular to the direction of the applied loads. Both positive and negative eccentricities must be considered for each direction of lateral load. If the building is in SDC C or higher and has a torsional or extreme torsional irregularity, the accidental torsion must be amplified per Section 12.8.4.3. It is possible that the torsion amplifier, given by Eq. 12.8-14, may be different at each level of the structure.

The third column of **Table G15–1** represents the orthogonal loading requirements, which are specified in Section 12.5. When orthogonal loading is required in SDC C and higher, it is permitted to satisfy these requirements by simultaneously applying 100 percent of the load in one direction (with torsion and torsion amplification if necessary) and simultaneously applying 30 percent of the orthogonal direction loading. Section 12.5.4 provides additional orthogonal loading requirements for certain interacting structural components in structures that have been assigned to SDC D though F. According to Section 12.8.4.2, the orthogonal direction load need not be applied with an eccentricity.

15.2 Load Combination Procedures Used in MRS Analysis

Where the modal response spectrum method of analysis is used, all signs in the member forces are lost because of the square root of the sum of the squares (SRSS) or complete quadratic combination (CQC) modal combinations. Additionally, it is common to apply accidental torsion as a static load and then combine this load with the results of the modal analysis. Orthogonal load effects may be handled in one of two manners:

1. Apply 100 percent of the spectrum in one direction, and run a separate analysis with 30 percent of the spectrum in the orthogonal direction. Member forces and displacements are obtained by SRSS or CQC for each analysis. Combine the two sets of results by direct addition.
2. Apply 100 percent of the spectrum independently in each of two orthogonal directions. Member forces and displacements are found by CQC. Combine to two sets of results by taking the SRSS of results from the two separate analyses.

The first method gives different results for different angles of attack for the main component of loading. The main advantage of the second method is that it produces the same results regardless of the angle of attack of the seismic loads (Wilson 2004). From either of these approaches, only two dynamic load analyses are required.

The results from the gravity and response spectrum analysis are then combined algebraically with the results of static torsion analyses, where the accidental torsion, amplified if necessary, is applied. There are only four basic cases of accidental torsion loading. These cases are illustrated in **Fig. G15–4**.

15.3 Special Seismic Load Combinations, Including the Overstrength Factor

In some cases, it might be necessary to design members or connections for load effects, including the overstrength factor, Ω_o. This factor is listed for each viable system in Table 12.2-1. The requirement to use the overstrength factor may come directly from ASCE 7, or it may come from the specification used to proportion and detail the member or connection. Examples of ASCE 7 requiring the use of the overstrength factor include elements supporting discontinuous walls or frames (Section 12.3.3.3) and the design of diaphragm collector elements (Section 12.10.2.1). An example from a material specification, in this case the *Seismic Provisions for Structural Steel Buildings* (AISC 2005a), is the requirement that moment connections in intermediate

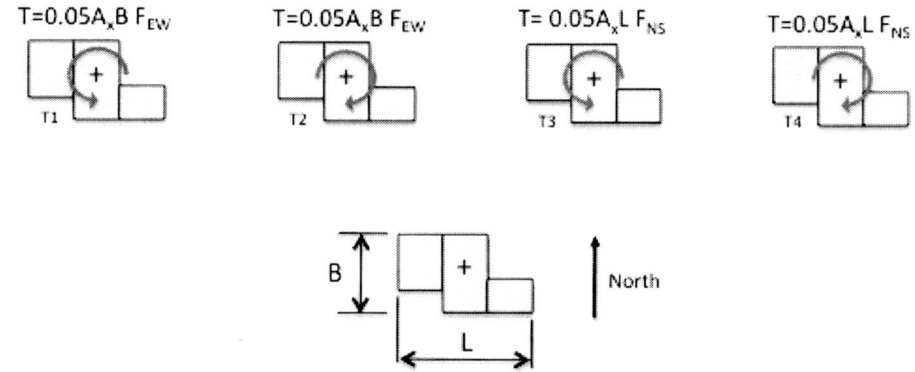

Figure G15-4 Static Load Cases for Torsion and MRS Analysis.

moment frames be designed for shear forces that are determined by "using load combinations stipulated by the applicable building code including the amplified seismic load." The amplified seismic load is defined as the "horizontal component of earthquake load E multiplied by Ω_o."

The special load combinations that include the overstrength factor are

Combination 5: $(1.2 + 0.2S_{DS})D + \Omega_o Q_E + 1.0L + 0.2S$
Combination 7: $(0.9 - 0.2S_{DS})D + \Omega_o Q^E + 1.6H$

where the only difference with respect to the standard load combination is that the term Ω_o replaces the redundancy factor ρ. Again, load combinations including the overstrength factor are required only for a few select members or connections. Most of the members and connections are designed using the standard load combination. There is no circumstance in which both the redundancy factor and the overstrength factor are used at the same time (for the same element or component).

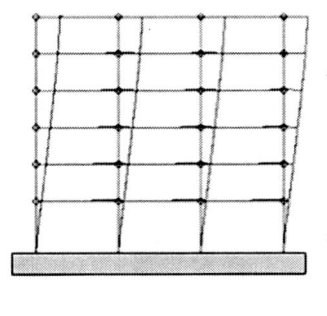

Example 16
Effective Seismic Weight (Mass)

In this example, the effective seismic weight is computed for an office and warehouse building in Burlington, Vermont. The example demonstrates the requirements for including both storage live load and snow load in the effective weight calculations. An additional example is provided to illustrate the computation of effective seismic weight for a one-story building with heavy self-supporting wall panels.

16.1 Four-Story Book Warehouse and Office Building in Burlington, Vermont

The building used for this example is an office and warehouse building in Burlington, Vermont. Plans and an elevation of the building are shown in **Fig. G16–1**. The first floor, at grade level, is used for both storage and office space, with about 70 percent of the area dedicated to storage. The second and third floors are used for storage only, and the fourth floor consists only of office space. The storage area is used primarily for boxes of textbooks to be used in the Burlington area public schools.

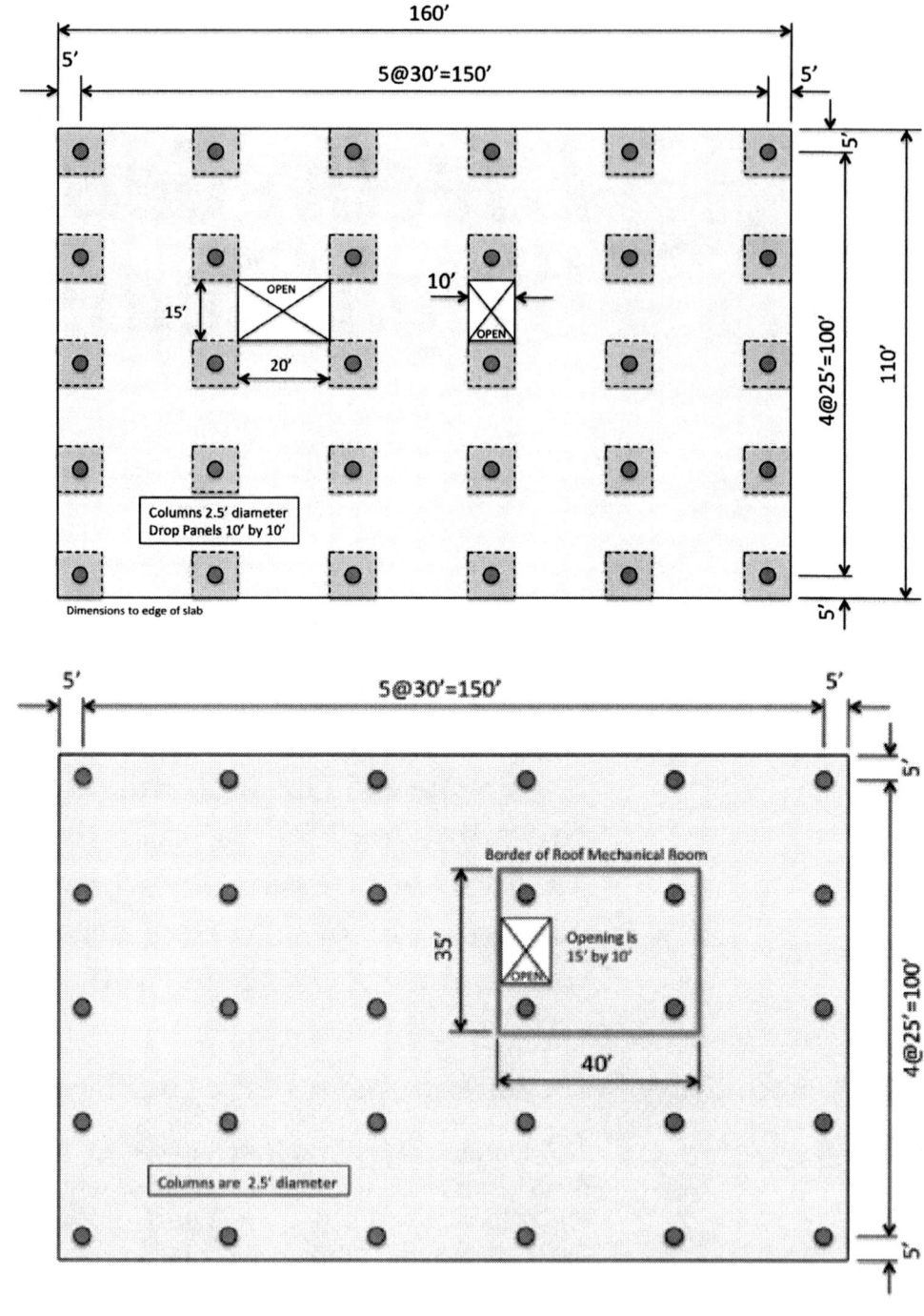

Figure G16–1 Plans and Elevation of Book Warehouse—(top) 2nd and 3rd Floor Plans, (bottom) Roof Plan (4th Floor Similar, Without Mechanical Room)

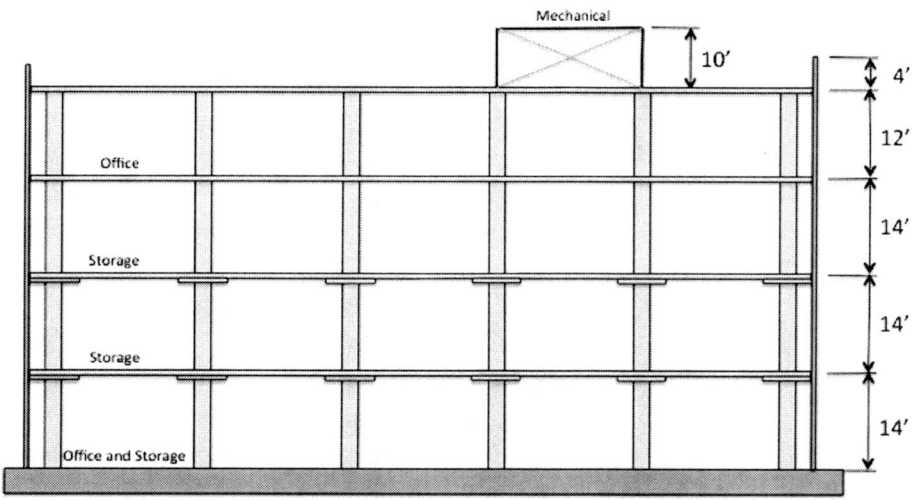

Figure G16–1   Plans and Elevation of Book Warehouse (continued)—Elevation

The structural system for the building is a prestressed concrete flat slab. This system supports gravity loads and acts as a moment resistant frame for both wind and seismic forces. For seismic design, the frame is classified as an ordinary moment resisting frame. It is expected that the equivalent lateral force (ELF) method would be used for the structural analysis of the system.

The slabs, constructed from lightweight structural concrete with a density of 115 lb/ft^3, have a basic thickness of 9 in. and are thickened to 12 in. at the second and third floor column areas to provide resistance to shear. The slabs at the fourth level (the office floor) and the roof do not have drop panels. A small mechanical penthouse, constructed from steel, is built over the roof slab, as shown in **Fig. G16–1**(roof plan and elevation). The penthouse is braced laterally and is sufficiently rigid to transfer its tributary roof and snow loads to the main roof level; hence, it is not considered a separate story.

For this example, it is assumed that the columns are also constructed from 115 lb/ft^3 lightweight concrete. In most cases, normal-weight concrete would be used for the columns of a concrete building.

The building is clad with lightweight precast concrete architectural panels with a thickness of 4 in. The concrete used for these panels has a density of 90 lb/ft^3. These panels have window openings that cover approximately 35 percent of the façade. These window areas weigh 12 lb/ft^2. The exterior wall extends 4 ft above the roof to form a solid parapet. The panels are supported vertically at grade and at levels 2, 3, and 4. The detailing of the panel connections is such that the panels are considered as effective seismic weight in each direction.

The books are stored in plastic containers, which in turn are supported by a steel rack system. The racks cover approximately 70 percent of the floor area. Small forklifts (not to be classified as permanent equipment)

are used to place and remove pallets of containers from the shelves. The design live load for the book storage area is 150 lb/ft². The rack storage system, which is anchored to the slab, weighs approximately 20 lb/ft². The racking system is laterally braced in two orthogonal directions with steel X-bracing. The system is sufficiently rigid to transfer the storage loads to the floor slabs.

The office area on the fourth floor of the building is designed for a live load of 50 lb/ft². A variety of work spaces are formed by a combination of fixed and movable partitions. A partition allowance of 15 lb/ft² is used in the design of the office floors. Two-thirds of this value, 10 lb/ft², is used for effective seismic weight as allowed by item 2 in Section 12.7.2.

The design dead load value used for the ceiling and mechanical areas of the main building is 15 lb/ft². Floor finishes in the office area are assumed to weigh 2.5 lb/ft². The floors in the storage areas are bare concrete.

The second and third floors have two openings, one (15 ft × 20 ft) to accommodate a hydraulic elevator for use in transporting the books (including the small forklifts), and the other (15 ft × 10 ft) for an elevator that services the offices on the fourth floor. The fourth floor and roof have only the smaller opening. Other minor openings exist in the floors and roof, but these openings are small and are not considered when computing the effective seismic weight. Two stairwells are also present in the building, but the weights of these are, on a pound per square foot basis, approximately the same as the floor slab.

The mechanical room contains a variety of heating, air conditioning, and ventilating equipment. The average dead load for the entire mechanical room, including the steel framing, roof, and equipment, is 60 lb/ft². The roofing over the remainder of the building (that area not covered by the mechanical room) is assumed to weigh 15 lb/ft².

The ground snow load for the building site is 60 lb/ft². Based on the procedures outlined in Chapter 7 of ASCE 7, it has been determined that the flat roof snow load is 42 lb/ft².

Calculation of the effective seismic weight, W, is based on the requirements of Section 12.7.2. The weight includes all dead load, a minimum of 25 percent of the floor live load in the storage areas, a 10 lb/ft² partition allowance where appropriate, total operating weight of permanent equipment, and 20 percent of the uniform design snow load when the flat roof snow load exceeds 20 lb/ft². Each of these load types is pertinent to the building under consideration.

16.1.1 Dead Load

The seismic load for the first floor level (a slab on grade) transfers directly into the foundation, so this load need not be considered as part of the effective seismic weight.

The loading for the second floor consists of the slab, drop panels, columns, storage rack system, ceiling and mechanical system, and exterior cladding.

Slab:
Total area = 160 × 110 − 15 × 20 − 15 × 10 = 17,150 ft^2
Unit weight = (9/12) × 115 = 86.2 lb/ft^2
Weight = 17,150 × 86.2/1,000 = 1,478 kips

Drop panels:
30 panels × 100 ft^2 per panel = 3,000 ft^2
Unit weight = (3/12) × 115 = 28.8 lb/ft^2
Weight = 3,000 × 28.8/1,000 = 86 kips

Columns:
Clear height tributary to first story = 13 ft
Clear height tributary to second story = 13 ft
Height tributary to second level = (13+13)/2 = 13 ft
Column area = 4.91 ft^2
Weight = 30 columns × 4.91 × 13 × 115/1,000 = 220 kips

Storage rack system:
Total area = 17,150 ft^2 (no deduction taken for columns)
Effective area = 0.7 × 17,150 = 12,005 ft^2
Unit weight = 20 lb/ft^2
Weight = 12,005 × 20/1,000 = 240 kips

Ceiling and mechanical system:
Total area = 17,150 ft^2 (no deduction taken for columns)
Unit weight = 15 lb/ft^2
Weight = 17,150 × 15/1,000 = 257 kips

Exterior cladding:
Perimeter = 2(160 + 110) = 540 ft
Height tributary to second level = 14 ft
Area of 4-in.-thick precast = 0.65 × 540 × 14 = 4,914 ft^2
Unit weight of panel = (4/12) × 90 = 30.0 lb/ft^2
Total panel weight = 4,914 × 30.0/1,000 = 147 kips
Area of glass windows = 0.35 × 540 × 14 = 2,646 ft^2
Unit weight of glass = 12 lb/ft^2
Total glass weight = 2,646 × 12/1,000 = 32 kips
Total cladding weight = 147 + 32 = 179 kips

Total dead load at second level = 1,478 + 86 + 220 + 240 + 257 + 179 = 2,460 kips.

The dead load on the third level is almost identical to that on the second level. The only difference is that the absence of the drop panels at the fourth story has a slight influence on the clear length of the columns at the third story. For this example, this small difference is ignored, and the same dead load is used for the second and third levels.

The dead load for the fourth level is computed as follows:

Slab: Total area = 160 × 110 − 15 × 10 = 17,450 ft^2
Unit weight = (9/12) × 115 = 86.2 lb/ft^2
Weight = 17,450 × 86.2/1,000 = 1,504 kips

Columns: Clear height tributary to third story = 13.25 ft
Clear height tributary to fourth story = 11.25 ft
Height tributary to second level = (13.25 + 11.25)/2 = 12.25 ft
Column area = 4.91 ft^2
Weight = 30 columns × 4.91 × 12.25 × 115/1,000 = 208 kips

Partitions: Total area = 17,450 ft^2 (no deduction taken for columns)
Unit weight = 10 lb/ft^2 (see Section 12.7.2, item 2)
Weight = 17,450 × 10/1,000 = 175 kips

Floor finish: Total area = 17,450 ft^2 (no deduction taken for columns)
Unit weight = 2.5 lb/ft^2
Weight = 17,450 × 2.5/1,000 = 44 kips

Ceiling and mechanical system: Total area = 17,450 ft^2 (no deduction taken for columns)
Unit weight = 15 lb/ft^2
Weight = 17,450 × 15/1,000 = 262 kips

Cladding: Perimeter = 2(160 + 110) = 540 ft
Height tributary to fourth level = 0.5(14 + 12) = 13 ft
Area of 4-in.-thick precast = 0.65 × 540 × 13 = 4,563 ft^2
Unit weight of panel = (4/12) × 90 = 30 lb/ft^2
Total panel weight = 4,563 × 30/1,000 = 137 kips
Area of glass windows = 0.35 × 540 × 13 = 2,457 ft^2
Unit weight of glass = 12 lb/ft^2
Total glass weight = 2,457 × 12/1,000 = 29 kips
Total cladding weight = 137 + 29 = 166 kips

Total dead load at fourth level = 1,504 + 208 + 175 + 44 + 262 + 166 = 2,359 kips.

The dead load for the roof level is computed as follows:

Slab: Total area = 160 × 110 − 15 × 10 = 17,450 ft^2
Unit weight = (9/12) × 115 = 86.2 ft^2
Weight = 17,450 × 86.2/1,000 = 1,504 kips

Columns: Clear height at fourth story = 11.25 ft
Height tributary to second level = (11.25)/2 = 5.62 ft
Column area = 4.91 ft^2
Weight = 30 columns × 4.91 × 5.62 × 115/1,000 = 95 kips

Ceiling and mechanical system:	Total area = 17,450 ft² (no deduction taken for columns)
	Unit weight = 15 lb/ft²
	Weight = 17,450 × 15/1,000 = 262 kips
Roofing:	Total area of main roof = 160 × 110 − 40 × 35 = 16,200 ft²
	Unit weight = 15 lb/ft²
	Weight = 16,200 × 15/1,000 = 243 kips
Mechanical area:	Total area = 40 × 35 = 1,400 ft²
	Unit weight = 60 lb/ft² (estimated)
	Weight = 1,400 × 60/1,000 = 84 kips
Cladding:	Perimeter = 2(160 + 110) = 540 ft
	Height tributary to roof = 6 ft
	Area of 4-in.-thick precast = 0.65 × 540 × 6 = 2,106 ft²
	Unit weight of precast = (4/12) × 90 = 30 lb/ft²
	Total panel weight = 2,106 × 30/1,000 = 63 kips
	Area of glass windows = 0.35 × 540 × 6 = 1,134 ft²
	Unit weight of glass = 12 lb/ft²
	Total glass weight = 1,134 × 12/1,000 = 14 kips
	Total cladding weight = 63 + 14 = 77 kips
Parapet:	Perimeter = 2(160 + 110) = 540 ft
	Height tributary to roof = 4 ft
	Area of 4-in.-thick precast = 540 × 4 = 2,160 ft²
	Unit weight of precast = (4/12) × 90 = 30 lb/ft²
	Total parapet weight = 2,160 × 30/1,000 = 65 kips

Total dead load at the roof = 1,504 + 95 + 262 + 243 + 84 + 77 + 65 = 2,330 kips.

The total dead weight for the building, including the mechanical level, is 9,609 kips. Using a building volume exclusive of the mechanical area of (160 × 110) × 54 = 950,400 ft³, the dead load density for the building is 9,609/950,400 = 0.0101 kips/ft³ or 10.0 lb/ft³. This weight is a bit heavier than that which would be appropriate for a low-rise office building, but it is reasonable for a concrete warehouse building. Calculation of building density is a good reality check on effective seismic weight. Low-rise buildings generally have a density in the range of 7 to 10 lb/ft³, depending on material and use.

16.1.2 Contribution from Storage Live Loads at Levels 2 and 3

As mentioned in the building description, the building has a design storage live load of 150 lb/ft². However, only 70 percent of each floor is reserved for storage, and the remainder is used for aisles, stairs, and restrooms. The openings for elevators are considered separately.

Building use statistics indicate that the storage racks are near capacity in the summer months when school is not in session and reduce to about 30 percent capacity during the fall and winter months. Section 12.7.2 states that a minimum of 25 percent of storage live load shall be used as effective seismic weight. For this facility, the 25 percent minimum is used. However, others might argue that, on the basis of use statistics, a larger portion of the load should be used[1].

The live load contribution to effective seismic weight is as follows for the second and third levels:

Total area = $160 \times 110 - 15 \times 20 - 15 \times 10 = 17{,}150$ ft^2
Effective area = $0.7 \times 17{,}150 = 12{,}005$ ft^2
Effective live load = $0.25(150) = 37.5$ lb/ft^2
Total live load contribution to seismic weight = $12{,}005 \times 37.5/1{,}000 =$ 450 kips

16.1.3 Contribution of Snow Load at Roof Level

Section 12.7.2 indicates that 20 percent of the uniform design snow load must be included in the effective seismic weight when the flat roof snow load exceeds 30 lb/ft^2. The flat roof snow load for this building is 42 lb/ft^2, so snow load must be included. The building has a flat roof (both the main roof and the mechanical room), so the uniform snow load is 42 lb/ft^2. Using a total area of $160 \times 110 = 17{,}600$ ft^2, the contribution from snow to the effective seismic weight is

Total area = $17{,}600$ ft^2
Effective snow load = $0.2(42) = 8.4$ lb/ft^2
Total snow load contribution to seismic weight = $17{,}600 \times 8.4/1{,}000 =$ 148 kips

The effective seismic weight for the entire system is summarized in **Table G16–1**. The design seismic base shear (Eq. 12.8-1) should be based on these weights, as should the distribution of forces along the height of the building (Eqs. 12.8-11 and 12.8-12). These forces should be placed at the center of mass of floors of the building, as appropriate. For this building, the center of mass is slightly offset from the plan center because of the floor openings and the somewhat eccentric location of the mechanical room.

The weights shown in **Table G16–1** are to be used in an ELF analysis of the system. If a three-dimensional modal analysis is used, the mass would need to be distributed to the various elements as necessary. When heavy cladding is used, it may be appropriate to include this cladding as line

---

1. The design of combined building-rack storage systems is considerably more complex than indicated in this example. See Chapters 13 and 15 of ASCE 7 for requirements for the design and attachment of rack systems to the building superstructure. See also the *Specifications for the Design, Testing, and Utilization of Industrial Steel Storage Racks* (RMI 2009)

masses, situated at the perimeter. If the mechanical penthouse covered more area, it might be appropriate to consider this as a separate level of the building. Section 12.2.3.1 covers situations when different lateral load resisting systems are used along the height of the building. The light rooftop structure used in this Seismic Design Category B example is exempt from the requirements of 12.2.3.1.

Consideration should also be given to the design, detailing, and anchorage of the steel rack system used in this building. Chapters 13 and 15 provide the requirements for the analysis and design of the system.

16.2 Low-Rise Industrial Building

In the previous example, the weight of the cladding parallel to and perpendicular to the direction of loading was included in the effective seismic weight in each direction. Thus, the effective seismic weight is the same in each direction.

For low-rise buildings, typically one story, the cladding panels may be detailed such that they are self-supporting when resisting seismic loads parallel to the plane of the panels and hence do not contribute to the seismic resistance of the main structural system when seismic loads act parallel to the plane of the panels. However, the effective weight of panels perpendicular to the direction of loading must be included.

Consider the low-rise industrial building shown in **Fig. G16–2**. The siding for the building consists of 5-in.-thick insulating concrete sandwich panels, which weigh 60 lb/ft². On the west face, there are openings covering approximately 35 percent of the wall panel area. There are only minor window and door openings on the other faces, and these openings are ignored in computing the effective seismic weight. Only the panel weight is considered herein.

For seismic forces in the north–south direction, the panel contribution to the effective seismic weight is based on one-half of the weight of the two 120-ft-long walls:

$$W_{panels, N-S} = 2 \times 120 \text{ ft} \times 18 \text{ ft} \times 60 \text{ lb/ft}^2 \times 0.5 = 129{,}600 \text{ lb} = 130 \text{ kips}$$

For seismic loads in the east–west direction, the effective seismic weight is

$$W_{panels, E-W} = (1 + 0.65) \times 50 \text{ ft} \times 18 \text{ ft} \times 60 \text{ lb/ft}^2 \times 0.5 = 44{,}550 \text{ lb} = 45 \text{ kips}$$

The 0.5 in the above calculations is based on half of the total effective seismic panel weight being carried by the steel framing and the remainder being resisted at the foundation.

Table G16–1 Summary of Effective Seismic Weight Calculation

Level	Load Contribution (kips)			
	Dead	Live	Snow	Total
2	2,460	450	0	2,910
3	2,460	450	0	2,910
4	2,359	0	0	2,359
R	2,330	0	148	2,478
Total	9,609	900	148	10,657

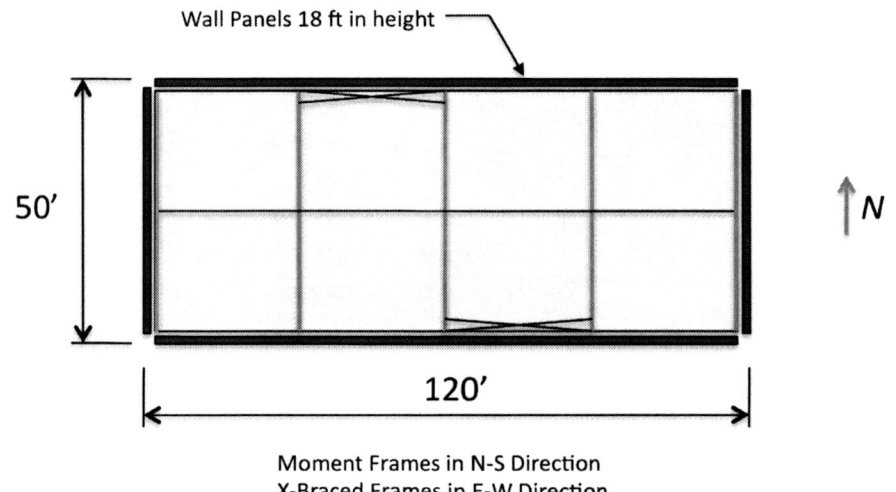

Moment Frames in N-S Direction
X-Braced Frames in E-W Direction

Figure G16–2 Plan View of a One-Story Industrial Building.

Example 17

Period of Vibration

This example explores the computation of the period of vibration of building structures. This example reviews the empirical methods that ASCE 7 provides for computing periods, computes periods for a few simple buildings, and then does a more detailed analysis wherein the empirical periods are compared to the period based on rational analysis.

17.1 Approximate Fundamental Period T_a

Section 12.8.2.1 addresses the computation of the approximate period of vibration of buildings. Three basic formulas are provided:

$$T_a = C_t h_n^x \qquad (12.8\text{-}7)$$

$$T_a = 0.1\,N \qquad (12.8\text{-}8)$$

$$T_a = \frac{0.0019}{\sqrt{C_w}} h_n \tag{12.8-9}$$

These formulas are highly empirical and are to be used for seismic analysis of building structures only. They are inappropriate for use in computing gust factors associated with wind loads (Section 6.5.8). Equation 12.8-7 applies to all buildings, Eq. 12.8-8 applies to certain moment frames, and Eq. 12.8-9 is applicable only for masonry or concrete shear wall structures. The primary use of T_a is in the computation of seismic base shear V. The period (in relation to T_S) is also used in determination of the appropriate method of analysis (Table 12.6-1).

In Eq. 12.8-7, the coefficient C_t and the exponent x come from Table 12.8-2 and depend on the structural system and structural material. These terms were developed from regression analysis of the measured periods of real buildings in California. The coefficients for buckling restrained braced (BRB) steel frames were inadvertently omitted from Table 12.8-2. The coefficients for BRBs are the same as those for eccentrically braced frames ($C_t = 0.03$, $x = 0.75$). These coefficients produce a somewhat longer period than those obtained for braced frames (all other systems in the table). This is logical because the bracing elements of BRB systems are always smaller than those required for traditional braced frames. Note also that the coefficients used for eccentrically braced steel frames are applicable only if the eccentrically braced frame is designed and detailed according to the requirements of the *Seismic Provisions for Structural Steel Buildings* (AISC 2005a).

When applying Eq. 12.8-7, the basic uncertainty is in the appropriate value to use for h_n, which is defined as "the height in ft above the base to the highest level of the structure." Section 11.2 of ASCE 7 defines the base as "The level at which the horizontal seismic ground motions are considered to be imparted to the structure." For a building on level ground without basements, the base may be taken as the grade level. In many cases, however, it may not be easy to establish an exact location of the base. This is particularly true when the building is constructed on a sloped site or when there are one or more basement levels. Also of some concern is the definition of the highest level of the structure. This height should not include small mechanical rooms or other minor rooftop appurtenances.

Consider, for example, the X-braced steel frame structures shown in **Fig. G17–1**. The first structure, (a), has no basement. Here, the height h_n is the distance from the grade level to the roof, not including the penthouse. Structure (b) is the same as structure (a) but has a full basement. At the grade level, the slab is thickened, and horizontal seismic force at the grade level is partially transferred though the diaphragms to exterior basement walls. Here again, the effective height should be taken as the distance from the grade level to the main roof. However, a computer model would produce a longer period for structure (b) relative to structure (a) because of

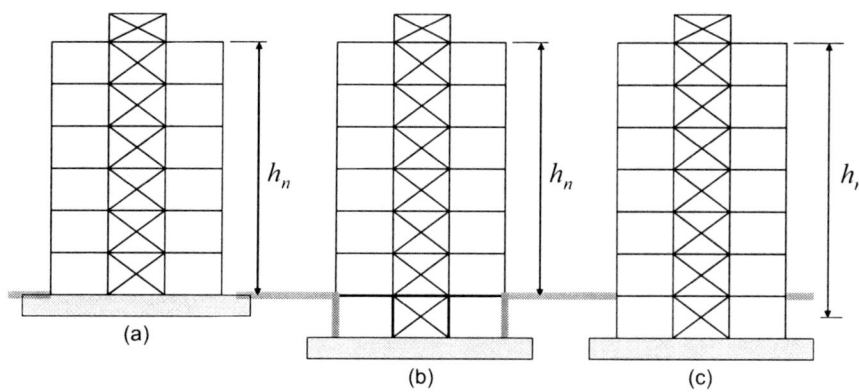

Figure G17–1 Finding the Effective Height h_n for a Braced Frame.

the axial deformations that occur in the subgrade columns of the braced frame of structure (b). In structure (c), the first grade slab is not thickened, and we do not expect braced frame shear forces to be completely transferred out through the first-floor diaphragm. However, some lateral support is provided by the soils adjacent to the basement walls. Here, the period of vibration is longer than for either (a) or (b) but may not be long as determined using the distance from the main roof to the top of the basement slab. Thus, a height from the main roof to the midlevel of the basement might be appropriate. Engineering judgment may be required. When in doubt, use the shortest reasonable height for h_n because this height produces the most conservative base shear.

Using structure (a) as an example, assume that the story height is 13 ft. Hence, $h_n = 13 \times 6 = 78$ ft. From Table 12.8-2, the "All other structural systems" classification applies, giving $C_t = 0.02$ and $x = 0.75$. Using Eq. 12.8-7,

$$T_a = C_t h_n^x = 0.02(78)^{0.75} = 0.525 \text{ s for the steel braced frame}$$

If a steel moment frame is used in lieu of the braced frame, Table 12.8-2 gives $C_t = 0.028$ and $x = 0.80$. Using Eq. 12.8-7,

$$T_a = C_t h_n^x = 0.028(78)^{0.8} = 0.914 \text{ s for the steel moment frame}$$

The longer period in this case reflects the fact that moment frames are generally more flexible than braced frames. Equation 12.8-8 is applicable for the moment frame and produces

$$T_a = 0.1 \, N = 0.1(6) = 0.60 \text{ s for the steel moment frame}$$

This shorter period is somewhat more conservative, which leads to a larger base shear than that given by Eq. 12.8-7.

17.1.1 Period T Used in the Equivalent Lateral Force Method

The total base shear and the distribution of lateral forces along the height of the structure are both functions of the period of vibration, T. To compute these quantities, the approximate period T_a may be used, giving $T = T_a$. This choice is generally conservative because periods computed on the basis of a rational structural analysis are almost always greater than those computed from the empirical formulas. In recognition of this fact, ASCE 7 (Section 12.8.2) allows the use of a modified period $T = C_u T_a$, where the coefficient C_u is provided in Table 12.8-1. The modifier may be used only if a period, called $T_{computed}$ herein, is available from a properly substantiated structural analysis.

Consider the braced frame structure (**Fig. G17–1(a)**) considered earlier, with $T_a = 0.525$ s. A computer analysis has predicted a period of $T_{computed} = 0.921$ s for the structure, so the modified period $T = C_u T_a$ may be used. Assuming that the structure is in a region of moderate seismicity with $S_{D1} = 0.25\,g$ and interpolating from Table 12.8-1,

$$C_u = 0.5(1.4 + 1.5) = 1.45$$

$$T = C_u T_a = 1.45 \times 0.525 = 0.761 \text{ s}$$

The period of 0.761 s must be used in the determination of the set of equivalent lateral forces from which the strength of the structure is to be evaluated, even though the rational period $T_{computed}$ was somewhat longer at 0.921 s. As explained in a separate example (Example 19), Section 12.8.6.2 of the standard allows the computed period to be used in the development of an alternate set of equivalent lateral forces that are used only for drift calculations.

17.1.2 What if the Computed Period $T_{computed}$ Is Less Than $T = C_u T_a$?

The computed period may turn out to be less than the upper limit period $C_u T_a$. Continuing with the braced frame, assume that the computed period $T_{computed} = 0.615 < C_u T_a = 0.761$ s. Although ASCE 7 is silent on this possibility, it is recommended that the lower period be used in the calculations. In the unlikely event that the computed period turned out to be less than $T_a = 0.525$ s, the period $T = T_a = 0.525$ s may be used because there is no requirement that a $T_{computed}$ shall be determined. If the computed period is significantly different from $C_u T_a$, say less than 0.5 $C_u T_a$ or more than 2 $C_u T_a$, the computer model should be carefully inspected for errors.

Table G17-1 summarizes the period values that should be used in the strength and drift calculations.

17.1.3 Period Computed Using Equivalent Lateral Forces and Resulting Displacements

Most commercial structural analysis computer programs can calculate the periods of vibration of building structures. If this capability is not available,

Table G17-1 Summary of Period Values to Be Used in Calculations

Situation	Period T to Be Used in Strength Calculations	Period T to Be Used in Drift Calculations
$T_{computed} \leq T_a$	T_a	T_a
$T_a < T_{computed} < C_u T_a$	$T_{computed}$	$T_{computed}$
$T_{computed} \geq C_u T_a$	$C_u T_a$	$T_{computed}$

the period can be estimated from the displacements produced from a set of lateral forces. The lateral force distributions provided by Eqs. 12.8-11 and 12.8-12 are well suited to this calculation, but these forces depend on the exponent k, which in turn depends on T, which we are trying to find. For the purpose of computing the lateral forces required for the period calculation, it is recommended that k be based on a trial period of $T = C_u T_a$.

The formula for computing the approximate period is based on a first-order Rayleigh analysis, and is as follows:

$$T = 2\pi \sqrt{\frac{\sum_{i=1}^{n} \delta_i^2 W_i}{g \sum_{i=1}^{n} \delta_i F_i}} \tag{G17-1}$$

where F_i is the lateral force at level i, δ_i is the lateral displacement at level i, W_i is the weight at level i, n is the number of levels, and g is the acceleration of gravity. When using the equation, the displacements δ are those that result directly from the application of the lateral forces F, and they do not include the deflection amplifier C_d.

The procedure is illustrated in **Table G17-2** for the braced frame in **Fig. G17-1**. For the example, it is assumed that the story weights are uniform at 1,100 kips per level. The mechanical penthouse is not included in the analysis. Using $T = C_u T_a = 0.761$ s, $k = 1.13$. The period computed from the information provided in **Table G17-2** is as follows:

$$T = 2\pi \sqrt{\frac{1}{386.4} \frac{39080}{2761}} = 1.20 \text{ sec}$$

17.1.4 Period Computed Using Computer Programs

The computed period depends directly on the assumptions made in modeling the mass and stiffness of the various components of the structure. In many cases, determination of the appropriate stiffness is not straightforward because these properties depend on a variety of factors, including the effective rigidity of connections, the degree of composite action, and the degree

Table G17–2 Determination of Period Using the Analytical Method

Level i	F_i (kips)	δ_i (in.)	W_i (kips)	$\delta_i F_i$ (kips-in.)	$\Sigma \delta_i^2 W_i$ (kips-in.²)
6	301	3.95	1,100	1,189	17,163
5	245	3.15	1,100	772	10,914
4	190	2.40	1,100	456	6,336
3	137	1.70	1,100	233	3,179
2	87	1.05	1,100	91	1,213
1	40	0.5	1,100	20	275
				S = 2,761	S = 39,080

of cracking in concrete. Although it is beyond the scope of this guide to provide detailed information on modeling, the following points are noted.

Deformations in the panel zones of the beam–column joints of steel moment frames are a significant source of flexibility in these frames. Two mechanical models for including such deformations are summarized in Charney and Marshall (2006). These methods are applicable to both elastic and inelastic systems. For elastic structures, it has been shown that the use of centerline analysis provides reasonable, but not always conservative, estimates of frame flexibility. Fully rigid end zones, as allowed by many computer programs, should never be used because they always result in an overestimation of lateral stiffness in steel moment resisting frames. The use of partially rigid end zones may be justified in certain cases, such as when doubler plates are used to reinforce the panel zone. Partially rigid end zones are also appropriate in the modeling of the joints of reinforced concrete buildings.

The effect of composite slabs on the stiffness of beams and girders may be warranted in some circumstances. When composite behavior is included, due consideration should be paid to the reduction in effective composite stiffness when portions of the slab are in tension (Schaffhausen and Wegmuller 1977; Liew 2001).

For reinforced concrete buildings, it is necessary to represent the effects of axial, flexural, and shear cracking in all structural components. Recommendations for computing cracked section properties may be found in Paulay and Priestly (1992) and other similar texts. In terms of the degree of cracking to use in the analysis, Section R10.11.1 of *Building Code Requirements for Structural Concrete* (ACI 2004) suggests that the period of vibration used to compute seismic base shear should based on the service load stresses. These stresses are presumably computed under gravity load conditions because this state would be the state of a new structure before an earthquake.

Shear deformations should be included in all structural analyses. Such deformations can be significant in steel moment frames. Additionally, shear deformations in reinforced concrete shear wall systems can actually dominate the flexibility when the walls have height-to-width ratios of less than 1.0.

17.2 Computing T_a for Masonry and Concrete Shear Wall Structures

Section 12.8.2.1 prescribes an alternate method for computing T_a for masonry or concrete shear wall structures. Two equations are provided:

$$T_a = \frac{0.0019}{\sqrt{C_W}} h_n \tag{12.8-9}$$

$$C_W = \frac{100}{A_B} \sum_{i=1}^{x} \left(\frac{h_n}{h_i}\right)^2 \frac{A_i}{\left[1 + 0.83\left(\frac{h_i}{D_i}\right)^2\right]} \tag{12.8-10}$$

where

A_B = area of the base of the structure (ft^2)
A_i = area of the web of wall i (ft^2)
D_i = plan length of wall i (ft)
h_i = height of wall i (ft)
x = number of walls in the direction under consideration

These equations are exercised using the structure shown in **Fig. G17–2**. The building has eight walls, with dimensions shown in the figure. The analysis is performed with four different assumptions on the height of the various walls. In all cases, the total building height is 120 ft and the area at the base of the building is 15,000 ft^2.

The results of the analysis are shown in **Table G17–3** together with the period computed using Eq. 12.8-7 and Table 12.8-2. As seen in the table, the period for the first system with staggered wall heights is less than the period of the same system with some of the walls extended to a greater height. The period for the system with all walls extended the full 120-ft height is more than three times that for the building with 40, 80, and 120-ft-tall walls. This result is highly counterintuitive because it would be expected that the added stiffness provided by extending the walls to a higher elevation would stiffen the structure and thereby reduce the period. It is therefore recommended that this method be used only with extreme caution.

17.3 Periods of Vibration for Three-Dimensional Systems

A computer program is required to determine the periods of vibration for 3D structures. The program reports periods (frequencies) for as many

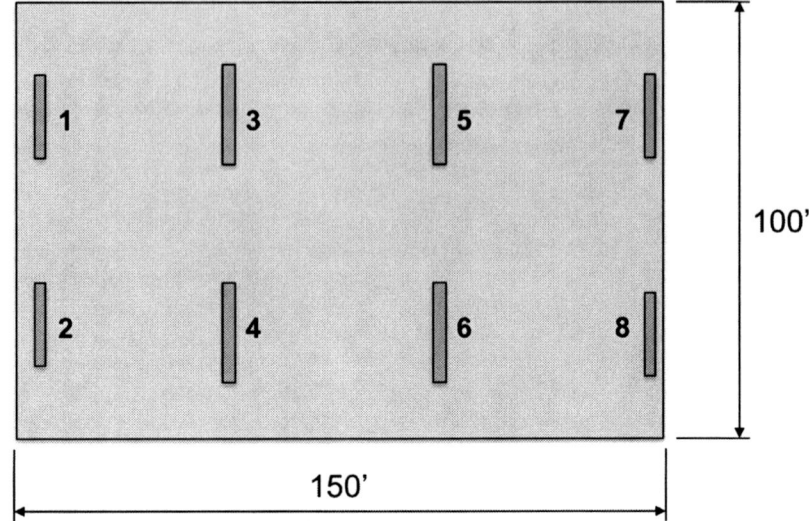

Figure G17–2 Concrete Shear Wall System Used for Period Determination.

Table G17–3 Computing the Period for Masonry and Concrete Shear Wall Structures

Height of Individual Walls (ft)			Using Eqs. 12.8-9 and 12.8-10		Using Eq. 12.8-7 with $C_t = 0.02$ and $x = 0.75$
1 and 2	3–6	7 and 8	C_W	T_a (s)	T_a (s)
40	120	80	0.134	0.622	0.725
80	120	80	0.0247	1.448	0.725
80	120	120	0.0184	1.678	0.725
120	120	120	0.0121	2.070	0.725

modes as the user requests. For regular rectangular buildings (regular in the sense of structural irregularities as described in Section 12.3.2), the first three modes generally represent the two orthogonal lateral modes and the torsional mode. These modes can occur in any order, and it is possible (although not generally desirable) that the first or second mode can be a torsional mode. It is also possible that the first mode (or any of the first several modes) can represent the vibration of a flexible portion of the system, such as a long cantilever, a long span beam, or even an error in connectivity. Thus, before using the periods in an analysis, the engineer should plot and animate the mode shapes to make sure that the proper periods are being used.

Interpretation and use of the modes and the periods of irregular buildings is often complex because even the first mode shape may represent a coupled lateral–torsional response. The period associated with the mode that has the highest mass participation factor in the direction under consideration should be used for the determination of the seismic forces in that direction. As with regular systems, the mode shapes should always be plotted and animated to make sure that they are reasonable. See Example 20 of this guide for more discussion on mode shapes and periods of vibration in highly irregular buildings.

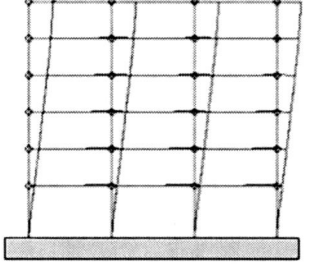

Example 18
Equivalent Lateral Force Analysis

In this example, an equivalent lateral force (ELF) analysis is performed for a reinforced concrete structure with intermediate moment frames resisting load in one direction and shear walls resisting the load in the other direction. Several aspects of the analysis are considered, including the determination of the design base shear, the distribution of lateral forces along the height, application of accidental torsion, and orthogonal load effects. Also included in a separate example is the demonstration of the use of a two-stage ELF procedure (Section 12.2.3.1), which may be applicable when the lower portion of the structure is significantly stiffer than the upper portion.

Before beginning the examples it is useful to review the pertinent equations used for ELF analysis. The first equation provides the design seismic base shear in the direction under consideration:

$$V = C_s W \qquad (12.8\text{-}1)$$

where W is the effective seismic weight of the building, and C_s is the seismic response coefficient. W is determined in accordance with Section 12.7.2, and C_s is calculated using one of five equations, numbered 12.8-2 through 12.8-6. The first three of these equations define an inelastic design response spectrum, and the last two provide lower limits on C_s.

The first three equations are as follows:

$$C_s = \frac{S_{DS}}{\left(\dfrac{R}{I}\right)} \quad \text{(applicable when } T \leq T_S\text{)} \qquad (12.8\text{-}2)$$

$$C_s = \frac{S_{D1}}{T\left(\dfrac{R}{I}\right)} \quad \text{(applicable when } T_S < T \leq T_L\text{)} \qquad (12.8\text{-}3)$$

$$C_s = \frac{S_{D1} T_L}{T^2\left(\dfrac{R}{I}\right)} \quad \text{(applicable when } T > T_L\text{)} \qquad (12.8\text{-}4)$$

The engineering units for C_s are in terms of the acceleration of gravity, g. This fact is not immediately clear from Eqs. 12.8-3 and 12.8-4 because both of these equations appear to have units of gravity per second. The apparent inconsistency in the units occurs because of the empirical nature of the equations. Equations 12.8-2 through 12.8-4 are plotted as an acceleration spectrum in **Fig. G18–1**. The specific values used to generate the figure are $S_S = 1.0$ g, $S_1 = 0.33$ g, Site Class B, $F_a = 1.0$, $F_v = 1.0$, $S_{DS} = 0.667$ g, $S_{D1} = 0.222$ g, $T_L = 4.0$ s, $I = 1$, and $R = 6$.

Two transitional periods separate the three branches of the response spectrum. These periods are determined as follows:

$$T_S = \frac{S_{D1}}{S_{DS}} \qquad \text{(Section 11.4.5)}$$

$$T_L \qquad \text{(See Figs. 22-15 to 22-20)}$$

T_S was derived by equating Eqs. 12.8-2 and 12.8-3 to find the period at which both branches of the spectrum produce the same value of C_s. Using the relationships provided in Chapter 11, it may be shown that T_S is the product of two ratios

$$T_S = \frac{S_1}{S_S} \times \frac{F_v}{F_a} \qquad (G18\text{-}1)$$

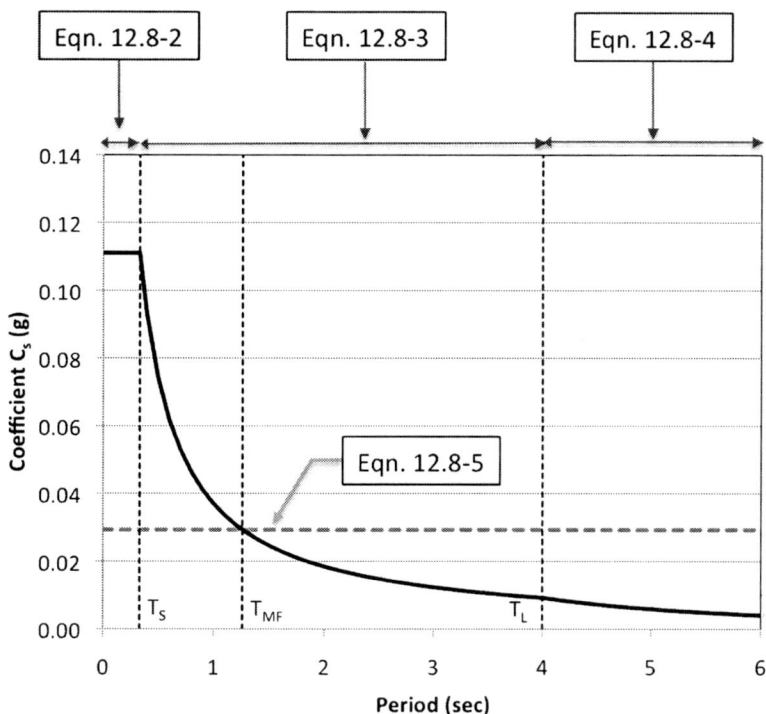

Figure G18–1 Inelastic Design Response Spectrum.

The ratio of S_1 to S_S varies across the United States, ranging from a low of about 0.2 (in New York City, for example), to a high of about 0.4 (in San Jose, for example). The ratio of F_v to F_a is a function of site class and seismicity. For Site Classes A and B, the ratio is fixed at 1.0. For higher site classes, the ratio increases with ground motion intensity and is a maximum of $2.4/0.9 = 2.67$ for Site Class E with $S_S > 1.25$ and $S_1 > 0.5$. Using the product of the ratios, it is possible for the transition period T_S to range from a low of about 0.2 s to a high of about 1.07 s.

The transitional period T_L comes from maps provided in Figs. 22-15 through 22-20. The minimum value of T_L is 4.0 s, and this value is applicable only in the Rocky Mountain region and the northwestern Hawaiian islands. Elsewhere, T_L ranges from 6.0 to 16.0 s.

Two equations are used to determine the minimum value of C_s:

$$C_s = 0.044\, S_{DS} I \geq 0.01 \quad \text{(applicable when } S_1 < 0.6\ g) \tag{12.8-5}$$

$$C_s \geq \frac{0.5 S_1}{\left(\dfrac{R}{I}\right)} \quad \text{(applicable when } S_1 \geq 0.6\ g) \tag{12.8-6}$$

Equation 12.8-5 is provided in Supplement No. 2 to ASCE 7-05 and supersedes the equation that was included in early printings of ASCE 7-05.

Equation 12.8-6 controls only in areas of high seismicity and is intended to account for the effects of near-field earthquakes. Equations 12.8-3 and 12.8-6, when set equal to each other, produce the period at which Eq. 12.8-6 controls C_s. This period, called T_{MN} herein, is

$$T_{MN} = 1.33\, F_v \tag{G18-2}$$

As seen in Table 11.4-2, when $S_1 > 0.5\,g$, F_v ranges from 0.8 for Site Class A to 2.4 for Site Class E, thus T_{MN} ranges from 1.06 s to 3.19 s. Clearly, Eq. 12.8-4 is not needed when Eq. 12.8-6 is applicable.

The period at which Eq. 12.8-5 controls may be determined by setting Eqs. 12.8-3 and 12.8-5 equal. This period, called T_{MF} herein, is

$$T_{MF} = \frac{22.7\, T_S}{R} \tag{G18-3}$$

This transitional period and a horizontal line representing Eq. 12.8-5 are shown in **Fig. G18-1**. The calculations are as follows:

$T_S = S_{D1}/S_{DS} = 0.222/0.667 = 0.3$ s

$T_{MF} = 22.7\, T_S/R = 22.7 \times 0.3/6 = 1.26$ s

$C_s\ (\text{minimum}) = 0.044 S_{DS} I = 0.044 \times 0.667 \times 1.0 = 0.029\, g$

Given the large range of values for T_S and R, it is difficult to develop a feel for T_{MF}. To help alleviate this problem, **Fig. G18–2** is provided, which plots the period values (vertical axis) versus the design short period acceleration S_{DS} (horizontal axis) at which Eq. 12.8-5 controls. The plots were developed for the case where S_1/S_S is 0.30. Somewhat different values can be obtained for other reasonable values of S_1/S_S. One graph is provided for each R value in the range 3 to 8, inclusive. For lower R values and firmer soils, Eq. 12.8-5 does not control until the period is greater than about 2.0 s. For low R values and softer soils (Classes D and E), the transition period is in the range of 3.0 s or higher. For higher R values and firm soils (Classes A and B), the transition period is in the range of 1.0 s, which is applicable for buildings in the range of 5 to 10 stories.

The basic conclusion from **Fig. G18–2** is this: The minimum base shear, given by Eqs. 12.8-1 and 12.8-5, is likely to control the magnitude of the seismic base shear in areas of moderate to high seismicity, even for low-rise buildings. This is true particularly when the building is situated on Site Class A or B soils.

The absolute lower limit on C_s is 0.01 g. This limit is used only in association with Eq. 12.8-5 (because Eq. 12.8-6 always results in a value greater than 0.01 g). This limit controls only when $S_{DS} < 0.227/I$. For Occupancy Categories I, II, or III, the importance factor is 1.0 or 1.25, giving limiting values of S_{DS} of 0.227 or 0.181 g, respectively. Thus, according to Table 11.6-1, the lower limit of 0.01 g is applicable only for Seismic Design Category A

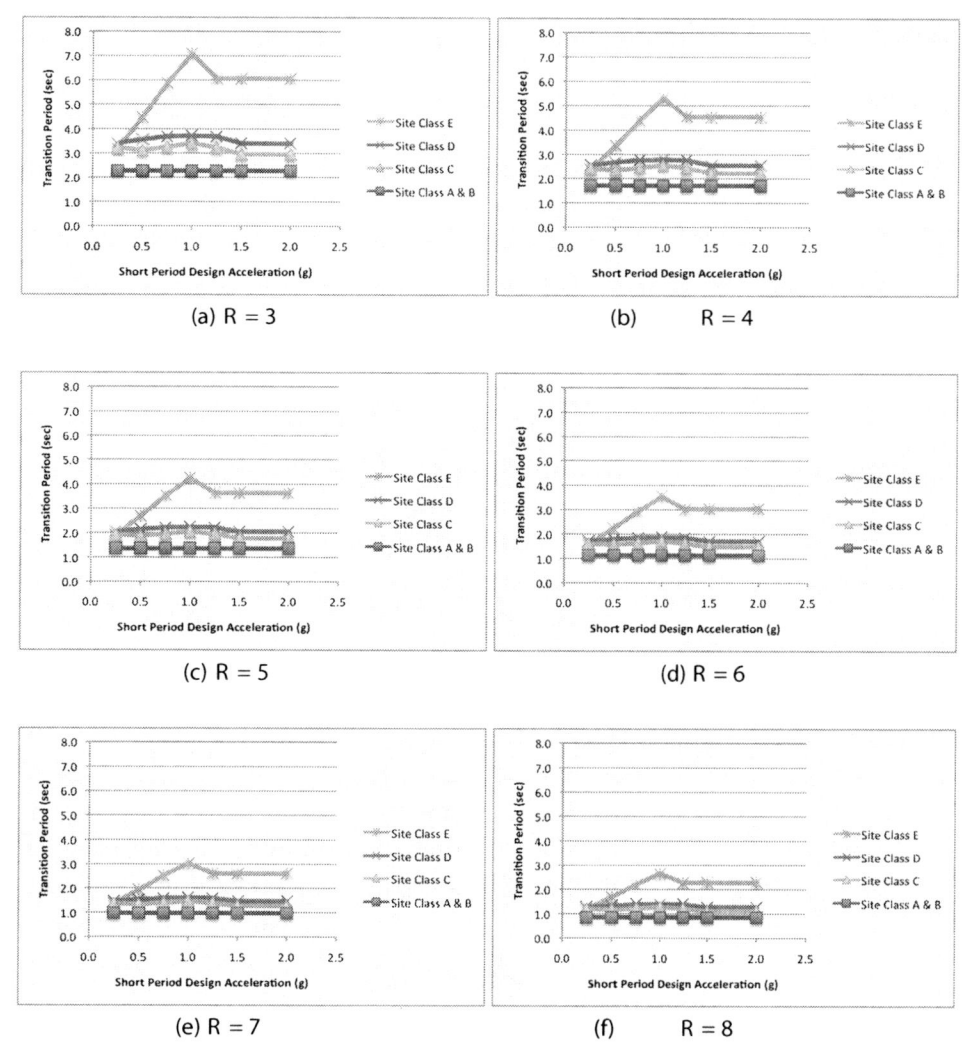

Figure G18–2 Period at Which Eq. 12.8-5 Controls C_s Value (Developed Using $S_1/S_S = 0.3$).

and B buildings. When the importance factor is 1.5, the limiting value of S_{DS} is 0.151 g, and for Occupancy Category IV, the Seismic Design Category is A. From this observation, the following statement may be made: The absolute minimum base shear, given by 0.01 W, may control, but only for Seismic Design Category B buildings and lower.

One more general statement may be made for **Fig. G18–2** (and similar figures for different S_1/S_S ratios, not shown herein): Equation 12.8-4 is never applicable for buildings on Site Class A and B soils and is rarely needed for buildings on Site Class C and D soils because the controlling period is never greater than 4.0 seconds (the minimum value of T_L).

18.1 Use of Equations 12.8-4, 12.8-5, and 12.8-6 in Computing Displacements

Equations 12.8-5 and 12.8-6 are intended to be used in determining design forces for proportioning members and connections but not for computing drifts that are to be used to check compliance with the drift limits specified in Section 12.12.1. Although it is not clearly stated in Section 12.8.6.2, the intent is that drift may be based on lateral forces developed using the computed period, with or without the upper limit $C_u T_a$ specified in Section 12.8.2. In no circumstances should drift be based on lateral forces computed using Eqs. 12.8-5 or 12.8-6[1] (i.e., the lower limits on C_s are not applicable to drift calculations).

The consequences of using Eq. 12.8-5 in drift computations are shown in the displacement spectra shown in **Fig. G18–3**. The displacement values include the deflection amplifier, $C_d = 5.5$. The solid line in the figure is the displacement spectrum obtained from Eqs. 12.8-2, 12.8-3, and 12.8-4. The dashed line is based on Eq. 12.8-5. The use of Eq. 12.8-5 produces highly exaggerated displacements at higher periods of vibration. A similar problem occurs when Eq. 12.8-6 controls.

18.2 Five-Story Reinforced Concrete Building

In this example, the equivalent lateral forces are determined for a five-story reinforced concrete building. The building is used to house a health care facility with a capacity of more than 50 patients, but no surgical facilities. Pertinent design information is summarized below:

Site Class = B
$S_S = 0.6\ g$
$S_1 = 0.18\ g$
$F_a = 1.0$ (from Table 11.4-1)
$F_v = 1.0$ (from Table 11.4-2)
$S_{DS} = (2/3)\ S_S \times F_a = 0.40\ g$ (11.4-1 and 11.4-3)
$S_{D1} = (2/3)\ S_1 \times F_v = 0.12\ g$ (11.4-2 and 11.4-4)
Occupancy Category = III (from Table 1-1)
Importance factor $I = 1.25$ (from Table 11.5-1)
Seismic Design Category = C (from Tables 11.6-1 and 11.6-2)

---

1. ASCE 7-10 will include a revision that states that lateral forces based on Eq. 12.8-5 need not be used for computing displacements, and that a separate set of forces based on the computed period (without the $C_u T_a$ limit) may be used for computing displacements. However, a similar proposal related to Eq. 12.8-6 was not approved. Hence, in ASCE 7-10, forces based on this Eq. 12.8-6 must also be used for drift calculations. For additional information on this subject, the reader is directed to a position statement developed by the Seismology Committee of the Structural Engineers Association of California (http://peer.berkeley.edu/tbi/news/applicability_of_UBC-97.html).

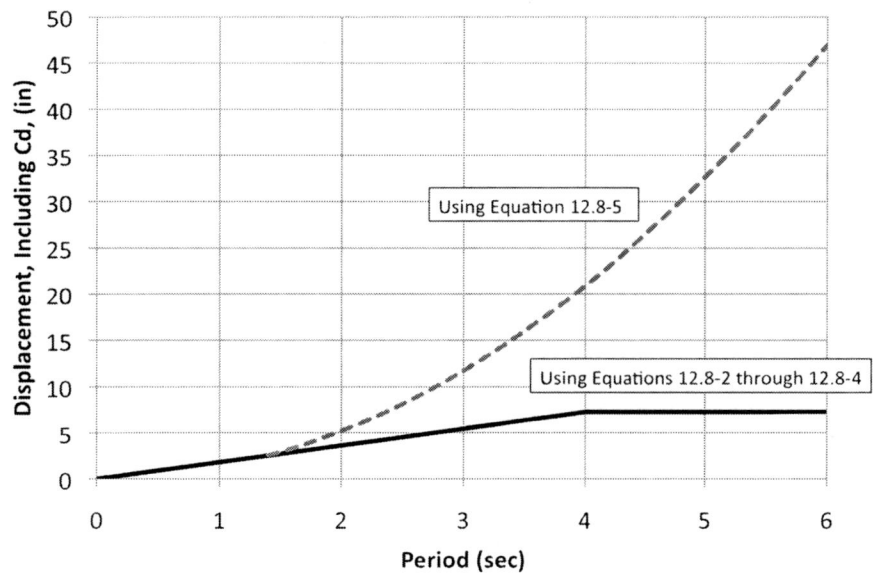

Figure G18–3 Displacement Response Spectra.

The ELF method of analysis is permitted for this SDC C building (Table 12.6-1). Based on the SDC of C, an ordinary reinforced concrete shear wall is allowed in the north–south direction and an intermediate reinforced concrete moment frame is allowed in the east–west direction (Table 12.2-1). Design parameters for these systems are as follows:

Ordinary Concrete Shear Wall

$R = 5$
$\Omega_o = 2.5$
$C_d = 4.5$
Height limit = None

Intermediate Concrete Moment Frame

$R = 5$
$\Omega_o = 3$
$C_d = 4.5$
Height limit = None

A plan of the building is shown in **Fig. G18–4**. The building has a 15-ft-deep basement, a 15-ft-tall first story, and 12-ft-tall upper stories. The slab at grade level is tied into exterior concrete basement walls, and these walls support the moment resisting frames. The base of the shear walls is interconnected by a grillage of foundation tie beams, and each individual wall is supported by reinforced concrete caissons that extend down to bedrock.

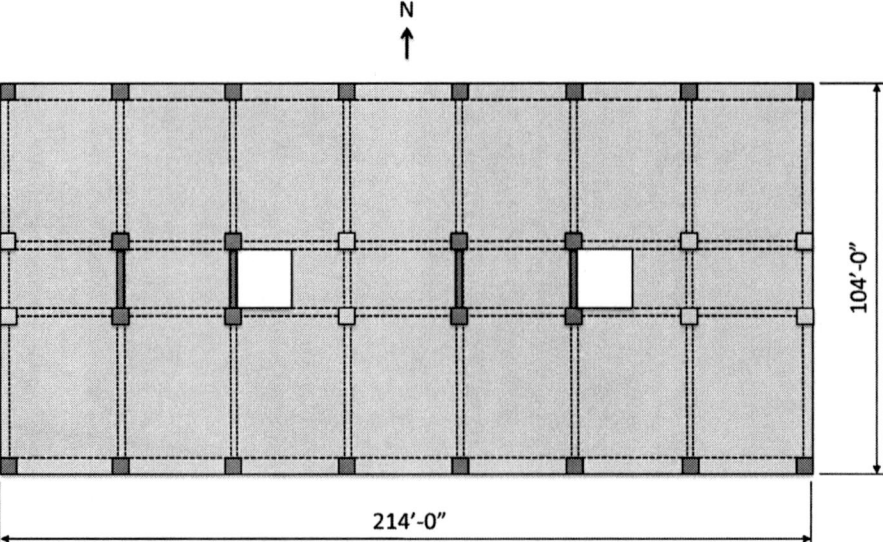

Figure G18–4 Plan View of a Five-Story Concrete Building.

Given this detailing, the base of the structure is at grade level, and the basement has no influence on the structural analysis.

The first step in the analysis is to determine the period of vibration. This period, which is different in the two directions because of the use of different structural systems, is calculated using Eq. 12.8-7:

$$T_a = C_t h_n^x \qquad (12.8\text{-}7)$$

For both directions, the height, h_n, is taken as the height above grade:

$$h_n = 15 + 4(12) = 63 \text{ ft}$$

For the shear wall system, Table 12.8-2 provides (in "All other structural systems") $C_t = 0.02$ and $x = 0.75$, giving

$$T_a = 0.02 \times 63^{0.75} = 0.45 \text{ s}$$

For the reinforced concrete moment frame, $C_t = 0.016$ and $x = 0.9$, and

$$T_a = 0.016 \times 63^{0.9} = 0.67 \text{ s}$$

At the preliminary design stage, the computed period from finite element analysis is not available. It is anticipated that these computed periods would exceed the approximate period, and hence, that the upper limit period $C_u T_a$ may be used. This assumption would need to be verified before proceeding with a final design.

Using Table 12.8-1 with $S_{D1} = 0.12\ g$, $C_u = 1.9 - 2S_{D1} = 1.9 - 2(0.12) = 1.66$. (See also Fig. GA-3, which provides interpolation equations for C_u). Thus the periods to be used for determination of ELF forces are

$T = 0.45 \times 1.66 = 0.75$ s for the shear wall system
$T = 0.67 \times 1.66 = 1.11$ s for the moment frame direction

The equivalent lateral forces are now developed for the shear wall direction in **Table G18–1**. These forces are based on the following effective seismic weight of the structure:

$W = 13{,}050\ k$

This weight is computed in accordance with Section 12.7.2.

The equation for finding C_s depends on the value of T, T_S, and T_{MF}.

$T_S = S_{D1}/S_{DS} = 0.30$ s (Section 11.4.5)
$T_{MF} = 22.7 T_s/R = 22.7 \times 0.3/5 = 1.36$ s (G18-1)

$T_S < T < T_{MF}$, so Eq. 12.8-3 controls. Using this equation gives

$$C_s = \frac{S_{D1}}{T\left(\dfrac{R}{I}\right)} = \frac{0.12}{0.75\left(\dfrac{5}{1.25}\right)} = 0.040 > 0.01$$

and from Eq. 12.8-1,

$V = C_s W = 0.040(13{,}050) = 522$ kips

The last item that is needed is the exponent k that is used in association with Eqs. 12.8-1 and 12.8-2 to determine the vertical distribution of forces. With T between 0.5 and 2.5 s, $k = 0.5T + 0.75 = 1.125$. (See Fig. GA-4, which provides interpolation functions for k).

Equation 12.8-3 also controls for the moment frame (east–west) direction. Thus, using $T = 1.11$ s,

$$C_s = \frac{S_{D1}}{T\left(\dfrac{R}{I}\right)} = \frac{0.12}{1.11\left(\dfrac{5}{1.25}\right)} = 0.027 > 0.01$$

$V = C_s W = 0.027(13{,}050) = 352$ kips

$k = 0.5T + 0.75 = 1.305$

The ELF story forces, shears, and overturning moments are shown in **Table G18–2**.

It is interesting to determine how the design forces might change if a special moment frame is used in the east–west direction instead of the

Table G18–1 ELF Forces, Shears, and Moments for North–South (Shear Wall) Direction

	1	2	3	4	5	6	7	8	9
	Story/ Level	H (ft)	h (ft)	W (kips)	Wh^k	$Wh^k/$ Total	F (kips)	Story Shear (kips)	Story OTM (ft-kips)
	5	12	63	2,500	264,361	0.328	171.4	171.4	2,056
	4	12	51	2,600	216,765	0.269	140.5	311.9	5,799
	3	12	39	2,600	160,295	0.199	103.9	415.8	10,788
	2	12	27	2,600	105,988	0.132	68.7	484.5	16,602
	1	15	15	2,750	57,868	0.072	37.5	522.0	24,432
	Totals	63	—	13,050	805,278	1.000	522	—	—

Table G18–2 ELF Forces, Shears, and Moments for East–West (Moment Frame) Direction

	1	2	3	4	5	6	7	8	9
	Story/ Level	H (ft)	h (ft)	W (kips)	Wh^k	$Wh^k/$ Total	F (kips)	Story Shear (kips)	Story OTM (ft-kips)
	5	12	63	2,500	557,288	0.350	123.1	123.1	1,478
	4	12	51	2,600	439,898	0.276	97.2	220.3	4,121
	3	12	39	2,600	309,965	0.195	68.5	288.8	7,587
	2	12	27	2,600	191,842	0.120	42.4	331.2	11,561
	1	15	15	2,670	94,218	0.059	20.8	352	16,841
	Totals	63	—	13,050	565,818	1.000	352	—	—

intermediate frame that was investigated previously. In this case, $R = 8$. The transitional period at which Eq. 12.8-5 controls is

$$T_{MF} = 22.7 T_s/R = 22.7 \times 0.3/8 = 0.85 \text{ s}$$

This period is less than the moment frame period of 1.11 s, and hence, Eq. 12.8-5 controls the base shear. This equation gives

$$C_s = 0.044 S_{DS} I = 0.044 \times 0.4 \times 1.25 = 0.022 > 0.01$$

Thus, the special moment frame with $R = 8$ is designed for $0.022/0.027 = 0.81$ times the design base shear for the $R = 5$ system, even though the apparent advantage of using $R = 8$ versus $R = 5$ is a design base shear ratio of 5/8 or 0.625. Given the extra costs of detailing a special moment frame compared to an intermediate moment frame, the economic incentive for using the special moment frame has likely disappeared.

18.2.1 Torsion, Amplification of Torsion, and Orthogonal Loading

Accidental torsion must be included in the structural analysis, per Section 12.8.4.2. It is likely that the building shown in **Fig. G18–4** has a torsional irregularity because of its rectangular shape and the interior location and nonsymmetric layout of the shear walls. If the torsional irregularity does exist, the accidental torsion must be amplified per Section 12.8.4.3, and 3D analysis must be used, as required by Section 12.7.3.

Section 12.5 establishes rules for direction of loading. Section 12.5.3, which pertains to SDC C buildings, requires that orthogonal loading effects be considered if the building has a Type 5 horizontal irregularity. The most practical method for including the orthogonal load requirements is to load the building with 100 percent of the load in one direction (including accidental torsion, amplified if necessary) and 30 percent of the load in the orthogonal direction. The 30 percent loading is applied without accidental torsion.

According to Section 12.13.4, the overturning forces imparted to the foundation by the shear walls may be reduced 25 percent. Although such a reduction may also be taken for the moment frame, the implementation of this reduction is not practical.

18.3 Two-Stage ELF Procedure per Section 12.2.3.1

Section 12.2.3.1 allows the use of a two-stage equivalent lateral force analysis for structures that have a flexible upper portion over a rigid lower portion, provided the following two criteria are met:

1. The stiffness of the lower portion must be at least 10 times the stiffness of the upper portion.
2. The period of the entire structure shall not be greater than 1.1 times the period of the upper portion considered as a separate structure fixed at the base.

When the criteria are met, the upper portion is analyzed as a separate structure using the appropriate values of R and ρ for the upper portion, and the lower portion is designed as a separate structure using the R and ρ for the lower portion. When analyzing the lower portion, the reactions from the upper portion must be applied as lateral loads at the top of the lower portion, and these reactions must be amplified by the ratio of the R/ρ of the upper portion to the R/ρ of the lower part of the lower portion. This amplification factor must not be less than 1.0.

The procedure is illustrated for the structure shown in **Fig. G18–5**. The base of the system is an ordinary precast shear wall ($R = 4$), and the upper portion is an ordinary concentrically braced steel frame ($R = 3.25$). The structure is assigned to Seismic Design Category B and thus has $\rho = 1$ (Section 12.3.4.1). The seismic weights at each level are shown in **Fig. G18–5**. The structure is situated on Site Class B soils. $S_{DS} = 0.25\ g$, and $S_{D1} = 0.072\ g$.

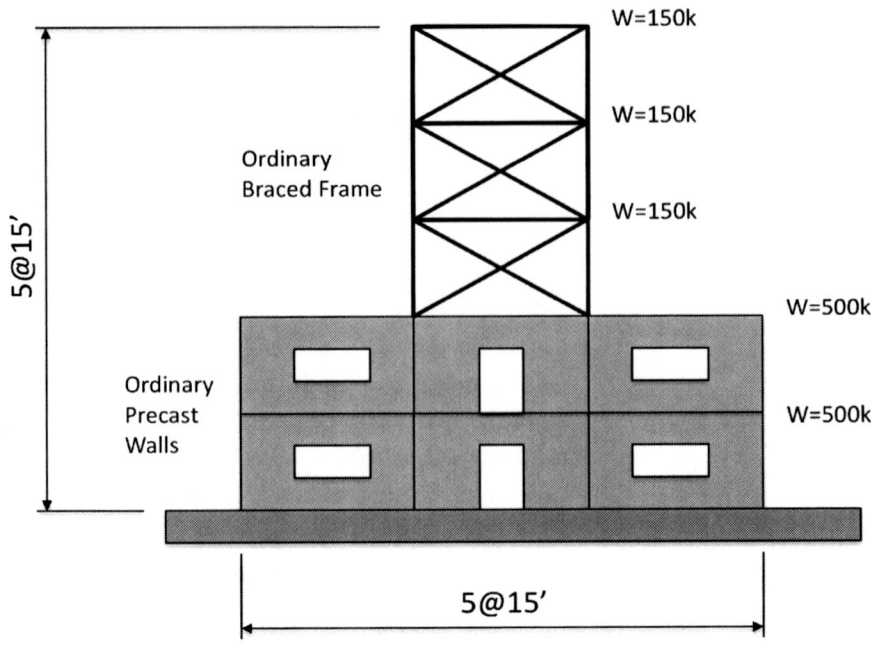

Figure G18–5   Structure Analyzed Using Two-Stage ELF Method.

The system was analyzed using SAP2000 (CSI 2009), wherein the precast walls were modeled with shell elements. The stiffness of the upper portion was determined by fixing it at its base and applying a 100-kips lateral load at the top. The displacement at the top was 0.607 in., thus the stiffness is approximately 100/0.607=165 kips/in. The stiffness of the lower portion was found by removing the upper portion and applying a 100-kips lateral force at the top of the lower portion. The displacement at the top of the lower portion was 0.0146 in., and the stiffness is 100/0.0146 = 6,872 kips/in. The ratio of the stiffness of the lower portion to the upper portion is 6,872/165 = 41.6, so the first criterion is met.

The periods of vibration were computed as follows: T for the whole system = 0.396 s, and T for the upper system fixed at its base = 0.367 s.

The ratio of the period of the entire structure to the period of the upper portion = 0.396/0.367 = 1.079, so the second criterion is met. Given that both criteria are met, the structure may be analyzed using the two-stage ELF method.

The periods of vibration used to determine the period ratios must be determined using a rational analysis, and not the approximate formulas provided by Section 12.8.2. Additionally, the stiffness of a structure of several degrees of freedom does not have a unique definition. The approach used above (wherein a 100-kips load was applied) is only one of a number of reasonable approaches that might be used.

The equivalent lateral forces for the upper and lower portions of the structure are shown in **Fig. G18–6**. The forces for the upper portion are

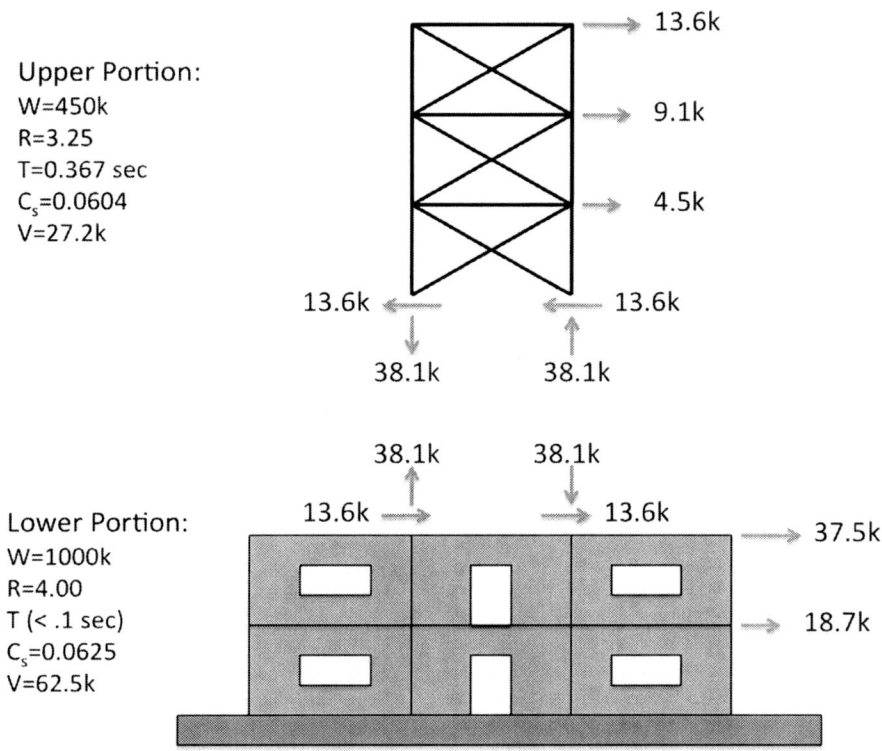

Figure G18-6 Forces on Upper and Lower Systems.

determined using $W = 450$ kips, $T = 0.367$ s, and $k = 1.0$. The approximate formula for period (including the $C_u T_a$ limit) is $1.7(0.02 \times 45^{0.75}) = 0.591$. This result is greater than the computed period (T = 0.367 s), so the computed period was used. The equivalent lateral forces for the lower portion were determined using $W = 1,000$ k and Eq. 12.8-2 (which always controls for very small T) and $k = 1.0$. When determining the reactions delivered from the upper portion to the lower portion, a magnification factor of 1.0 was used because the R for the upper portion is less than the R for the lower portion

When checking drift, the C_d values appropriate for the upper or lower portion should be used. The drift check is not shown herein.

Example 19

Drift and P-Delta Effects

In this example, the drift for a nine-story office building is calculated in accordance with Section 12.8.6 and then checked against the acceptance criteria provided in Section 12.12 and specifically in Table 12.12-1. P-delta effects are then reviewed in accordance with Section 12.8.7. The building is analyzed using the equivalent lateral force method, but only those aspects of the analysis that are pertinent to drift and P-delta effects are shown.

The seismic force resisting structural system for the building is a structural steel moment resisting space frame, placed at the perimeter. Each perimeter frame has five bays, each 30 ft wide. There is one 12-ft-deep basement level, an 18-ft-high first story, and eight additional upper stories, each with a height of 13 ft. The building has no horizontal or vertical structural irregularities. The total height of the structure above grade is 122 ft. The building is located in Seattle, Washington, on Site Class D soils. An elevation of one of the perimeter frames is shown in **Fig. G19–1**. The girder sizes are shown in the figure because these are needed in the P-delta check.

In this example, the lateral loads used to compute drift are determined two ways. In the first case, loads are based on the upper limit period of vibration, $T = C_u T_a$, computed in accordance with Section 12.8.2. In the second case, lateral loads are computed using the period of vibration determined from a rigorous (finite element) analysis of the system. This period is referred to as $T_{computed}$ in this example. The second case is used to illustrate the benefit of Section 12.8.6.2, which allows $T_{computed}$ to be used in the determination of the lateral loads that are applied to the structure for the purpose of computing drift.

The design spectral accelerations for the Site Class D location are as follows:

$$S_S = 1.25\ g \qquad F_a = 1.0 \qquad S_{DS} = 0.83\ g$$
$$S_1 = 0.5\ g \qquad F_v = 1.5 \qquad S_{D1} = 0.50\ g$$

The Occupancy Category for the building is II, the importance factor is 1.0, and the Seismic Design Category is D.

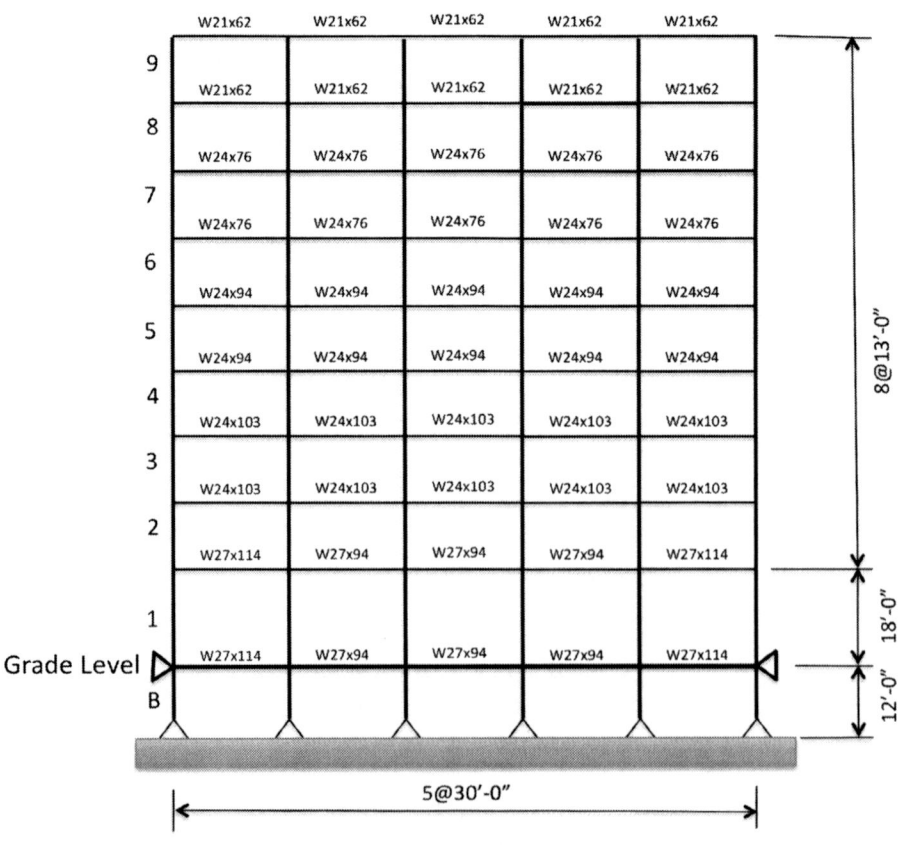

Figure G19–1 Elevation of Building Used for P-Delta Analysis.

Using Table 12.2-1, the response modification coefficient, R, for the special steel moment frame is 8, and the deflection amplification factor, C_d, is 5.5. There is no height limit for special moment frames.

The effective seismic weight of the building, W, is 21,000 kips. The approximate fundamental period of the building is computed using Eq. 12.8-7, $h_n = 122$ ft, and coefficients $C_t = 0.028$ and $x = 0.8$ from Table 12.8-2:

$$T_a = C_t h_n^x = 0.028 \times 122^{0.8} = 1.31 \text{ s}$$

The height used in the above computation (122 ft) is based on the assumption that the structure is laterally restrained at the grade level (as shown by a pin support in **Fig. G19–1**).

Using Table 12.8-1, the coefficient for upper limit on period, C_u, is 1.4, thus the period T used for determining base shear and lateral loads is

$$T = C_u T_a = 1.4 \times 1.31 = 1.83 \text{ s}$$

However, the upper limit on period may be used only if a computed period based on a properly substantiated structural analysis is available. Such an analysis was performed, resulting in $T_{\text{computed}} = 3.20$ s. This period was computed using centerline analysis, which approximately accounts for deformations in the panel zones of the beam–column joints. The inclusion of panel zone effects is required in Section 12.7.3. P-delta effects were not included in the analysis because these effects are considered separately, as shown later in this example.

The computed period is significantly greater than the adjusted period $T = C_u T_a$, and this difference is of some concern. Because of this concern, the computer model was thoroughly checked, and no errors were found. However, differences between the empirical and the computed period, with the computed period greater than the empirical period, are not unusual in moment frame analysis. As shown later in this example, however, periods that are significantly greater than the $C_u T_a$ upper limit may be an indicator that the building is too flexible.

19.1 Drift Computations Based on $T = C_u T_a = 1.83$ s

The design seismic base shear based on the upper limit for period of vibration is computed using Eq. 12.8-1:

$$V = C_s W \tag{12.8-1}$$

Equation 12.8-3 controls the value of C_s. Using $T = C_u T_a = 1.83$ s,

$$C_s = \frac{S_{D1}}{T(R/I)} = \frac{0.5}{1.83 \times (8/1)} = 0.0341 \text{ g}$$

The minimum values of C_s, based on Eq. 12.8-5, need not be checked here because the base shear is being determined for use in a drift analysis, and not a strength (member proportioning and detailing) analysis.[1]

Using $W = 21,000$ kips for the entire building,

$$V = C_s W = 0.0341 \times 21,000 = 716 \text{ kips}$$

The equivalent lateral force (ELF) story forces for the full building (not a single frame) are shown in column 4 of **Table G19–1**. These forces were computed according to Eqs. 12.8-11 and 12.8-12, with the exponent $k = 1.665$ for $T = 1.83$ s. Application of these forces to the building resulted in the story displacements δ_{xe} shown in column 5 of the table. See Eq. 12.8-15 and Fig. 12.8-2 of ASCE 7 for a description of symbols used in computing drift.

Interstory drifts (the displacement at the top of a story minus the displacement at the bottom of the same story) are shown in column 6 of **Table G19–1**. The design-level interstory drifts, Δ_x, computed according to Eq. 12.8-15, are shown in column 7. These drifts are based on $C_d = 5.5$ and $I = 1.0$.

According to Table 12.12-1, the interstory drift limits for this Occupancy Category II building are 0.02 times the story height. These values are

Table G19–1 Drift Analysis of Building of Fig. G19-1 Using Limiting Period $T = C_u T_a$

1	2	3	4	5	6	7	8	9	10
Story x	Height (in.)	W_x (kips)	F_x (kips)	δ_{xe} (in.)	Δ_{xe} (in.)	$\Delta_x = C_d \Delta_{xe}/I$ (in.)	Limit (in.)	Ratio	Okay?
9	156	2,250	173.2	6.471	0.448	2.46	3.12	0.79	OK
8	156	2,325	148.3	6.023	0.683	3.76	3.12	1.20	NG
7	156	2,325	120.1	5.340	0.713	3.92	3.12	1.26	NG
6	156	2,325	94.2	4.627	0.804	4.42	3.12	1.42	NG
5	156	2,325	71.0	3.823	0.784	4.31	3.12	1.38	NG
4	156	2,325	50.4	3.039	0.827	4.55	3.12	1.46	NG
3	156	2,325	32.8	2.212	0.749	4.12	3.12	1.32	NG
2	156	2,325	18.3	1.463	0.692	3.81	3.12	1.22	NG
1	216	2,475	7.9	0.771	0.771	4.24	4.32	0.98	OK

1. It is the author's belief that Section 12.8.6.2 may be used to justify this point, although this is not the original intent of 12.8.6.2. Section 12.8.6.2 allows drift to be computed on the basis of the computed period of vibration, without the upper limit $C_u T_a$. Please see Example 18 for a more detailed discussion.

shown in column 8 of **Table G19–1**. The ratio of the design-level drift to the drift limit is provided in column 9. The limits are exceeded (ratios greater than 1.0) at all stories except for story 1 and story 9. In story 4, the computed drift is 1.46 times the specified limit.

19.2 Drift Computations Based on $T = T_{computed} = 3.2$ s

The calculations are now repeated for the same structure analyzed with lateral forces consistent with the computed period of 3.2 s.

The seismic coefficient C_s for $T = 3.2$ s is

$$C_s = \frac{S_{D1}}{T(R/I)} = \frac{0.5}{3.2 \times (8/1)} = 0.0195g$$

and the base shear for determining drift is

$$V = C_s W = 0.0195 \times 21{,}000 = 409 \text{ kips}$$

The results of the drift analysis are presented in **Table G19–2**. The exponent k for computing the distribution of lateral forces is 2.0 in this case. The drifts have reduced substantially and do not exceed the limiting values at any level. Thus, in this case, there appears to be a significant advantage to using the computed period when calculating drifts.

Table G19–2 Drift Analysis of Building of Fig. G19-1 Using $T_{computed}$

1	2	3	4	5	6	7	8	9	10
Story x	Height (in.)	W_x (k)	F_x (k)	δ_{xc} (in.)	Δ_{xc} (in.)	$\Delta_x = C_d \Delta_{xc}/I$ (in.)	Limit (in.)	Ratio	Okay?
9	156	2,250	109.5	3.849	0.285	1.59	3.12	0.51	OK
8	156	2,325	90.4	3.564	0.419	2.31	3.12	0.74	OK
7	156	2,325	70.1	3.145	0.442	2.42	3.12	0.78	OK
6	156	2,325	52.4	2.703	0.481	2.64	3.12	0.85	OK
5	156	2,325	37.3	2.222	0.468	2.59	3.12	0.83	OK
4	156	2,325	24.7	1.754	0.478	2.69	3.12	0.84	OK
3	156	2,325	14.7	1.276	0.438	2.43	3.12	0.78	OK
2	156	2,325	7.3	0.838	0.397	2.18	3.12	0.70	OK
1	216	2,475	2.6	0.441	0.441	2.43	4.32	0.56	OK

However, a drift analysis is not complete without performing a P-delta check. This check is performed in the following section for the case where $T_{computed}$ is used to determine the story forces used in drift analysis. (The P-delta analysis using the loads and displacements based on $T = C_u T_a$ is similar to that using $T = T_{computed}$. The only difference in the results is based on the difference in the coefficient k used to compute the distribution of vertical forces along the height of the building.)

19.3 P-Delta Effects

The P-delta check is carried out in accordance with Section 12.8.7. In the P-delta check, the stability ratio is computed for each story, in accordance with Eq. 12.8-16:

$$\theta = \frac{P_x \Delta I}{V_x h_{sx} C_d} \tag{12.8-16}$$

where

- P_x is the total vertical design gravity load at level x.
- Δ is the interstory drift at level x and is based on center of mass story displacements computed using Eq. 12.8-15 and thus includes the deflection amplifier C_d but does not include the importance factor.
- I is the importance factor.
- V_x is the total design shear at level x and is based on C_s computed using Eq. 12.8-3 and thus includes the importance factor as a multiplier.
- h_{sx} is the story height.
- C_d is the deflection amplifier from Table 12.2-1.

Equation 12.8-16 as shown above includes the importance factor I in the numerator. This factor was erroneously omitted in ASCE 7-05, but it must be present to offset the importance factor that is included in the story shears. Hence, the I term that appears explicitly in the numerator and the I term that is included in V_x cancel out. Also, the term C_d, which is included in the numerator as part of Δ and appears explicitly in the denominator, also cancels out in Eq. 12.8-16. In essence, therefore, the stability ratio, if properly computed, is independent of I and C_d (one should get the same value regardless of which I and C_d are used).

The results of the P-delta analysis are shown in **Table G19–3.** Column 3 of the table provides the individual story dead loads, and column 4 contains the fully reduced story live loads. The accumulated story gravity forces, P, are given in column 6 of the table. The gravity forces are unfactored, in accordance with the definition of P_x in Section 12.8.7. The shears in column 7 are the accumulated story shears. The interstory drifts that are given in column 8 include the C_d amplifier. The calculated stability ratios are given in column 9 of the table, where the maximum value of 0.133 occurs at level 2.

The limiting value of θ is given by Eq. 12.8-17:

$$\theta_{max} = \frac{0.5}{\beta C_d}$$

where β is the ratio of shear demand to shear capacity of the story, which in essence is the inverse of the story overstrength. If β is taken as 1.0, the limit on θ is 0.5/5.5 = 0.091 and is the same for all stories. It can be seen from **Table G19–3** that the ratio of 0.091 is exceeded at stories 1 through 6.

It is likely that the story overstrength is greater than 1.0 because of the strong-column, weak-beam rules that are built into the various design specifications, such as the *Seismic Provisions for Structural Steel Buildings* (AISC 2005a). (Many other factors would also contribute to overstrength, such as actual versus nominal yield strength, strain hardening, and plastic hinging sequence). Column 11 of **Table G19–3** shows the value of $1/\beta$ that would be required to make the limiting value of θ at each story equal to the computed value (as listed in column 9 of **Table G19–3**). For example, for the second level, we want

$$\theta_{max} = 0.133 = \frac{0.5}{C_d \beta} = \frac{0.5}{5.5} \times \frac{1}{\beta}$$

Now, solving for $1/\beta$,

$$\frac{1}{\beta} = \frac{C_d \theta_{max}}{0.5} = \frac{5.5 \times 0.133}{0.5} = 1.47$$

If it is assumed that at least this level of overstrength can be provided at level 2 and that the required overstrengths can also be provided at the remaining levels, the drifts can be updated to include P-delta effects, and they can be compared again to the limiting values. The revised drift at each level is the drift without P-delta divided by the quantity $(1 - \theta)$, where θ is taken from column 9 of **Table G19–3**. The results of the calculations are shown in column 12. These drifts are compared to the drift limits (column 13). These limits are based on the "All other structures" classification in Table 12.12-1. In all cases, the drifts, including P-delta, are less than the specified limit.

19.4 Computation of Actual Story Overstrengths

In the above computations, it has been assumed that the overstrengths required to keep θ below or equal to θ_{max} could be obtained. The calculation

Table G19–3 P-Delta Effects

1	2	3	4	5	6	7	8	9	10	11	12	13	14
Story	Height (in.)	P_{dead} (k)	P_{live} (k)	p_{total} (k)	P_{total} (k)	V (k)	D (in.)	θ	θ Limit Using $\beta = 1$	Required $1/\beta$	Revised Drift (in.)	Limit (in.)	Ratio 2/3
9	156	2,150	450	2,600	2,600	109	1.59	0.044	0.091	0.48	1.67	3.12	0.535
8	156	2,225	450	2,675	5,275	200	2.31	0.071	0.091	0.78	2.49	3.12	0.797
7	156	2,225	450	2,675	7,950	270	2.42	0.083	0.091	0.91	2.64	3.12	0.846
6	156	2,225	450	2,675	10,625	322	2.64	0.101	0.091	1.11	2.94	3.12	0.942
5	156	2,225	450	2,675	13,300	360	2.59	0.111	0.091	1.22	2.91	3.12	0.932
4	156	2,225	450	2,675	15,975	384	2.59	0.125	0.091	1.38	2.96	3.12	0.947
3	156	2,225	450	2,675	18,650	399	2.43	0.132	0.091	1.46	2.80	3.12	0.898
2	156	2,225	450	2,675	21,325	406	2.18	0.133	0.091	1.47	2.52	3.12	0.808
1	216	2,375	450	2,825	24,150	409	2.43	0.120	0.091	1.32	2.76	4.32	0.638

of actual story strengths is not straightforward and would typically require a series of nonlinear static analyses. A simplified method for estimating story strengths is provided by Section C3 of the commentary to the AISC *Seismic Design Manual* (AISC 2005b). If the structure has been designed in accordance with the strong-column, weak-beam design rules, the plastic story strength may be estimated from the following equation:

$$V_{yi} = \frac{2\sum_{j=1}^{n} M_{pGj}}{H} \quad \text{(AISC Seismic Specification Commentary Eq. C3-2)}$$

where M_{pGj} is the plastic moment capacity of the girder in bay j, n is the number of bays, and H is the story height under consideration.

Using the section sizes shown in **Fig. G19–1** and assuming a yield stress of 50 kips/in.^{2} for steel, the story strengths for one frame are computed as shown in column 2 of **Table G19–4**. Column 3 of the same table lists the strength demands, which are based on the story force values in column 4 of **Table G19–1**, but divided by 2.0 to represent a single frame.

Before one calculates the overstrength, one must divide the values in column 3 by the quantity $(1 - \theta)$, as required by Section 12.8.7. This step is done for all values, even when the stability ratio θ is less than 0.10. The ratio of the computed capacity to the strength demand is shown in column 5 of **Table G19–4**. Clearly, the ratios all exceed the required ratios ($1/\beta$ values) shown in column 11 of **Table G19–3**.

The AISC formula does not work for braced frames, dual systems, or any other type of structure except a moment frame. Calculating story strengths

Table G19–4 Estimated Story Overstrength Factors

1	2	3	4	5
Story	V_y (kips)	V_{demand} (kips)	$V_{demand}/(1-\theta)$ (kips)	Ratio 2/4
9	461	86	90	5.09
8	461	160	173	2.66
7	641	220	240	2.66
6	641	267	298	2.14
5	814	303	341	2.38
4	814	328	376	2.16
3	897	344	396	2.26
2	897	354	408	2.19
1	704	358	407	1.73

for general structural systems is not straightforward and may not even be possible without a detailed nonlinear static pushover analysis.

Future versions of ASCE 7 are likely to abandon Eq. 12.8-17 in favor of requiring the designer to demonstrate stability through the use of a nonlinear static pushover analysis whenever the stability ratio for any level is greater than 0.1.

19.5 Backcalculation of Stability Ratios When P-Delta Effects Are Included in Analysis

Many structural analysis programs provide the option to directly include P-delta effects. If an analysis is run with and without P-delta effects, the story stability ratios may be backcalculated from the results of the two analyses. Although in theory the approach may be used for three-dimensional analysis, the most straightforward use is for two-dimensional analysis, and this approach is demonstrated here.

This method is based on the following equation:

$$\Delta_f = \frac{\Delta_0}{1 - \frac{P\Delta_0}{VH}} = \frac{\Delta_0}{1-\theta} \tag{G19-1}$$

where

Δ_f = the story drift from the analysis including P-delta effects,
Δ_0 = the drift in the same story for the analysis without P-delta effects,
P = vertical design load in the story (the same as that used in Eq. 12.8-16),
V = the seismic story shear (the same as that used in Eq. 12.8-16), and
H = the story height (using the same length units as those used for drift).

The story drifts must be computed from the same lateral loads that produce the story shears. A rearrangement of terms in Eq. G19-2 produces the simple relationship

$$\theta = 1 - \frac{\Delta_0}{\Delta_f} \tag{G19-2}$$

Equation G19-2 is demonstrated through the use of data provided in **Tables G19–2** and **G19–3** and from a separate analysis that included P-delta effects. The results are provided in **Table G19–5** The stability ratios are similar to those provided in column 9 of **Table G19–3**.

As a final note, when Eq. 12.8-16 is used, the displacements should not include P-delta effects. Doing so would include P-delta effects twice: once in the original analysis, and once again when computing the stability ratios.

Table G19–5 Stability Ratios Backcalculated from Analysis Including P-Delta Effects

1	2	3	4
Story	Δ_0 (in.)	Δ_1 (in.)	θ
9	0.285	0.299	0.047
8	0.419	0.450	0.069
7	0.442	0.481	0.081
6	0.481	0.537	0.104
5	0.468	0.525	0.109
4	0.478	0.548	0.128
3	0.438	0.505	0.133
2	0.397	0.460	0.137
1	0.441	0.502	0.122

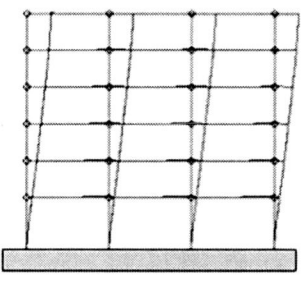

Example 20
Modal Response Spectrum Analysis

In this example, the modal response spectrum (MRS) method of analysis is used to analyze a six-story moment resisting frame. The analysis is performed in accordance with Section 12.9 of ASCE 7. The results from the MRS analysis are compared to the results obtained from an equivalent lateral force (ELF) analysis of the same frame. In most cases, an MRS analysis is carried out using commercial finite element analysis software. For this example, however, the analysis is performed using a set of Mathcad routines that have been developed by the author. The use of Mathcad provides details of the analysis that are not easily extracted from the commercial software.

The building analyzed in this example is located in Savannah, Georgia. The site for this building is the same as that used in Examples 4 and 5. The building is six stories tall and is used for business offices. According to the descriptions in Table 1-1, the Occupancy Category is II, and from Table 11.5-1, the importance factor I is 1.0. Pertinent ground motion parameters are summarized below:

Site Class = D
S_{DS} = 0.395 g
S_{D1} = 0.188 g

Given the above parameters and an Occupancy Category of II, the building is assigned to Seismic Design Category C. The structural system for the building is an intermediate steel moment frame. According to Table 12.2-1, such systems are allowed in Seismic Design Category C, and they have no height limit. The relevant design parameters for the building are

R = 4.5
C_d = 4

Plan and elevation drawings of the building are shown in **Fig. G20–1**. Moment resisting frames are placed along lines 1 and 6 in the east–west direction and on lines C and E in the north–south direction. The frames that resist loads in the east–west direction have a series of setbacks, as shown in the elevation. The frames that resist loads in the north–south direction do not have setbacks. This example considers only the analysis of the frame with setbacks, and more specifically, the frame on gridline 1.

Because of the setbacks, the structure has both a weight irregularity and a vertical geometric irregularity, as described in Table 12.3-2. A preliminary analysis of the complete structure indicates that the structure does not have a torsional irregularity.

According to Table 12.6-1, the equivalent lateral force method may be used to analyze this Seismic Design Category C building. However, because of the vertical irregularities, the modal response spectrum approach is used instead. The details of the MRS method are presented in this example, and the results are compared to those obtained from an ELF analysis of the same system.

The story heights and seismic weights for the frame are provided in **Table G20–1**. The weights represent the effective seismic mass that is resisted by the frame on gridline 1 only, and they thus represent one-half of the mass of the building.

20.1 Modeling of Structural System

Section 12.7.3 requires that a three-dimensional model be used whenever the structure has a Type 1a (torsional) or Type 1b (extreme torsional) horizontal irregularity. Such an irregularity does not exist in this case, so it is

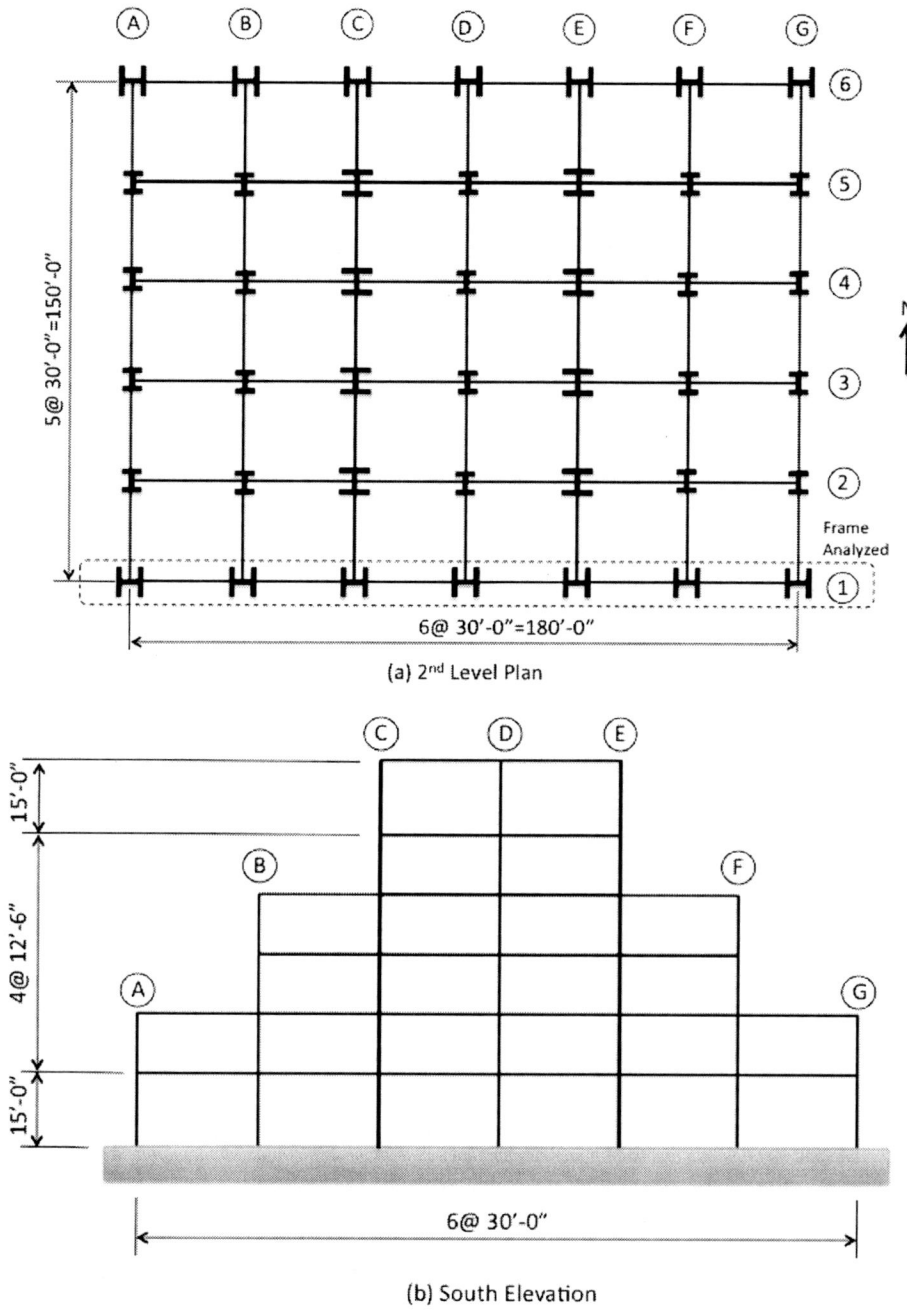

Figure 20–1 Plan and Elevations of Building Analyzed Using MRS.

acceptable to analyze the structure separately in each direction. Therefore, a planar analysis model was used. Note, however, that accidental torsion must be considered and would typically be applied using a static analysis. The torsional analysis is not provided in this example.

For this six-story frame, there are 37 nodes in the analytical model. The columns are assumed to be fixed at their base, leaving 30 unrestrained

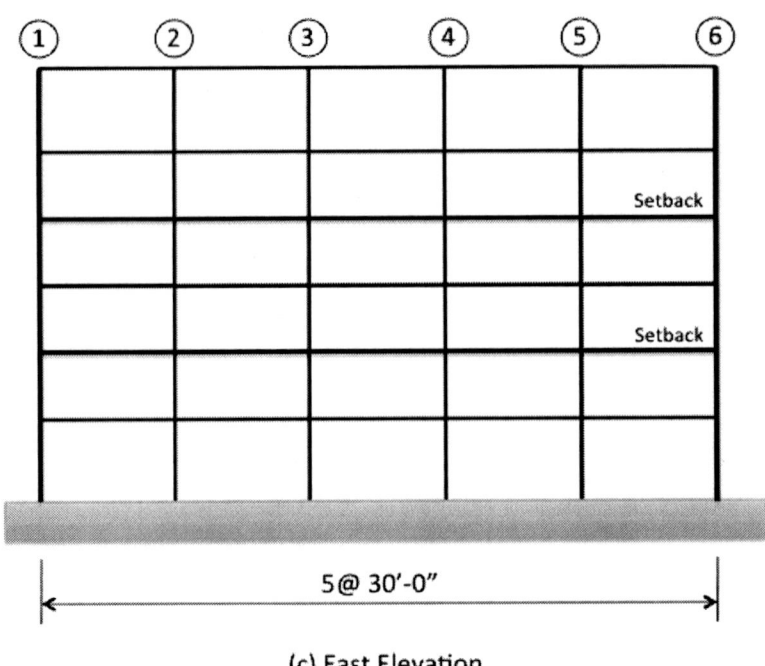

(c) East Elevation

Figure G20-1   Plan and Elevations of Building Analyzed Using MRS. *(continued)*

Table G20-1   Story Properties for Moment Frame

Story	Height (ft)	Seismic Weight[a] (kips)	Seismic Mass[a] (kips-s^2/in.)
6	15.0	500	1.294
5	12.5	525	1.359
4	12.5	1,000	2.588
3	12.5	1,025	2.653
2	12.5	1,500	3.882
1	15.0	1,525	3.947
Total	80.0	6,075	15.722

nodes. With three degrees per node (two translations and a rotation) there are 90 degrees of freedom (DOF) in the frame. If it is further assumed that the floor diaphragms are rigid in their own plane, the number of DOF reduces to 66. (Thirty nodes with two degrees of freedom (one vertical and one rotational), plus six lateral DOF, one for each story.)

For dynamic analysis, only six degrees of freedom need to be considered, where each degree of freedom represents the lateral displacement at a floor level. As shown in **Fig. G20–2**. these DOF are numbered from the top down. However, the computer model used in the static analysis includes

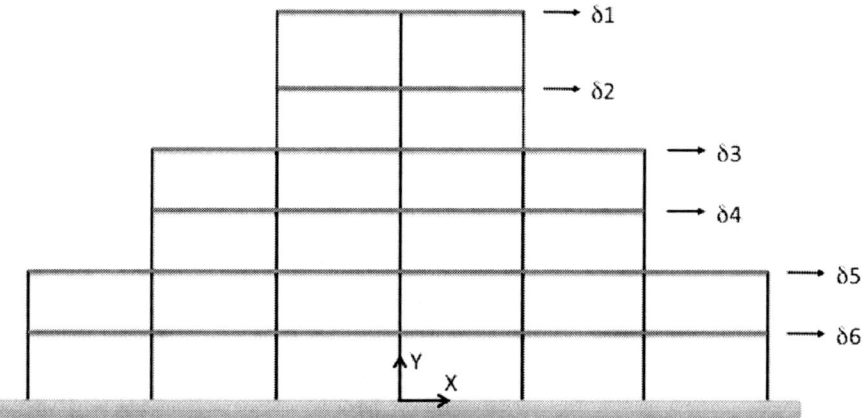

Figure G20-2 Degrees of Freedom Used in MRS Analysis.

all 66 DOF, and the system's stiffness matrix was assembled on this basis. A 6-DOF dynamic stiffness matrix K was obtained from the larger matrix by static condensation. Units of all terms are kips per inch.

$$K = \begin{bmatrix} 128.72 & -192.46 & 72.98 & -10.48 & 1.44 & -0.22 \\ -192.46 & 521.94 & -410.20 & 91.89 & -12.92 & 1.99 \\ 72.98 & -410.20 & 885.79 & -714.80 & 191.61 & -28.39 \\ -10.48 & 91.89 & -714.80 & 1352.60 & -896.57 & 198.49 \\ 1.44 & -12.92 & 191.61 & -896.57 & 1822.86 & -1374.71 \\ -0.22 & 1.99 & -28.39 & 198.49 & -1374.71 & 2260.02 \end{bmatrix}$$

A lumped mass idealization is used, resulting in a diagonal mass matrix. This matrix, M, shown below, is based on the story weights shown in **Table G20-1**. The units of the terms of the mass matrix are kips-s^2/in.

$$M = \begin{bmatrix} 1.29 & 0 & 0 & 0 & 0 & 0 \\ 0 & 1.36 & 0 & 0 & 0 & 0 \\ 0 & 0 & 2.59 & 0 & 0 & 0 \\ 0 & 0 & 0 & 2.65 & 0 & 0 \\ 0 & 0 & 0 & 0 & 3.88 & 0 \\ 0 & 0 & 0 & 0 & 0 & 3.95 \end{bmatrix}$$

20.2 Determination of Modal Properties

The next step in the analysis is to determine the modal properties for the system. These properties include

1. The modal frequencies, ω, and periods, T
2. The mode shapes, ϕ

3. The modal participation factors, Γ
4. The effective modal mass, m

The mode shapes, ϕ, and modal frequencies, ω, are obtained by solving the eigenvalue problem

$$K\phi = \omega^2 M\phi \tag{G20-1}$$

This equation has six solutions, one mode shape vector ϕ, and one circular vibration frequency ω for each dynamic degree of freedom in the system. Mathcad was used to determine the mode shapes and frequencies, with the results shown below. The individual mode shapes are stored columnwise in Φ, and the individual circular frequencies (in units of radians per second) are stored along the diagonal of Ω. The mode shapes are normalized such that the maximum ordinate in each mode is exactly 1.0. The mode shapes carry no physical units. The first three modes are plotted in **Fig. G20–3**.

$$\Phi = \begin{bmatrix} \phi_1 & \phi_2 & \phi_3 & \phi_4 & \phi_5 & \phi_6 \end{bmatrix} \begin{bmatrix} 1.000 & 1.000 & -0.948 & -0.453 & -0.329 & -0.108 \\ 0.804 & 0.234 & 0.654 & 1.000 & 1.000 & 0.405 \\ 0.615 & -0.300 & 1.000 & -0.052 & -0.777 & -0.591 \\ 0.444 & -0.449 & 0.058 & -0.599 & 0.494 & 1.000 \\ 0.264 & -0.410 & -0.798 & 0.081 & 0.314 & -0.922 \\ 0.131 & -0.235 & -0.640 & 0.384 & -0.640 & 0.781 \end{bmatrix}$$

$$\Omega = \begin{bmatrix} 3.359 & & & & & \\ & 7.143 & & & & \\ & & 11.996 & & & \\ & & & 20.574 & & \\ & & & & 26.236 & \\ & & & & & 32.458 \end{bmatrix}$$

The modal participation factors for each mode i are computed as follows:

$$\Gamma_i = \frac{\phi_i^T M R}{\phi_i^T M \phi_i} \tag{G20-2}$$

where the T superscript on ϕ indicates a matrix transpose, and the modal excitation vector R is a one-column matrix containing a value of 1.0 in each row.

The effective modal masses are then obtained from the following equation:

$$m_i = \Gamma_i^2 \phi_i^T M \phi_i \tag{G20-3}$$

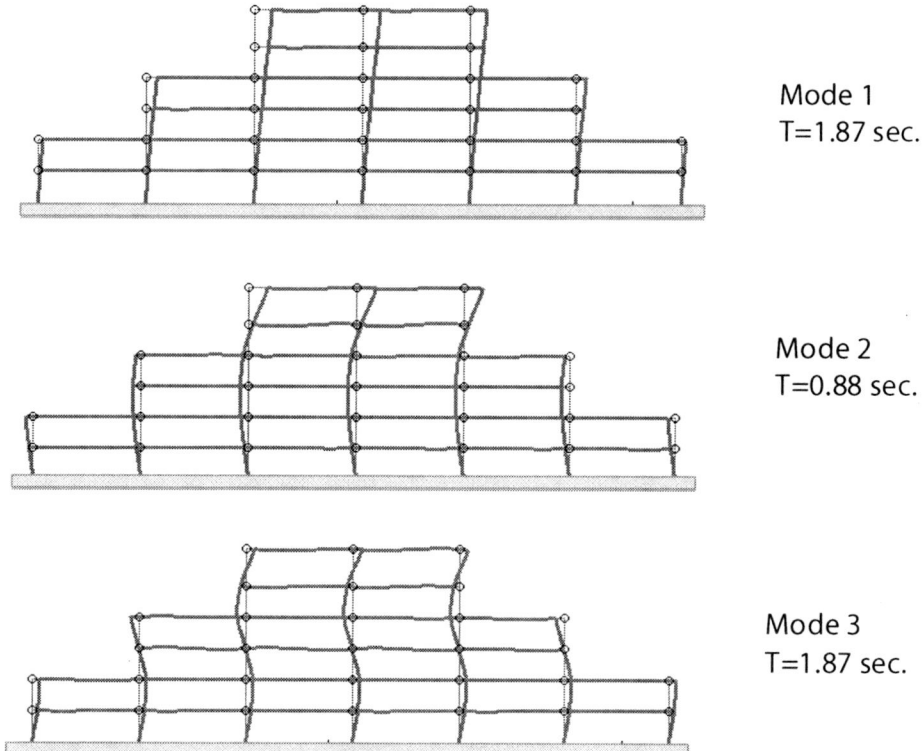

Figure G20–3 Mode shapes for First Three Modes.

The modal properties are summarized in **Table G20–2**. (These properties are typically reported by commercial software.) Column 3 contains the modal periods, which are each equal to $2\pi/\omega$. Column 6 contains the accumulated effective modal masses. When all six modes are included, the accumulated mass is 15.723 kips-s²/in., which as required by theory, is equal to the total mass in the system. Column 7 contains the accumulated mass divided by the total mass, represented as a percent. Section 12.9.1 of ASCE 7 requires that an MRS analysis must include enough modes to capture at least 90 percent of the actual mass in each orthogonal direction. Only three modes need be considered for the current analysis because the accumulated effective mass for the first three modes is greater than 90 percent of the total mass. For brevity, the example proceeds with three modes. However, for a system with only 6 dynamic DOF, it would be more reasonable to include all modes in the analysis.

20.3 Development of Elastic Response Spectrum

The loading for the system is based on the acceleration response spectrum defined in Section 11.4.5. The spectrum is plotted in **Fig. G20–4** for S_{DS} = 0.395 g and S_{D1} = 0.188 g. The acceleration values associated with all six of the modal periods of vibration are listed in columns 3 and 4 of **Table G20–3**. (All six modes are shown for completeness, even though only the first three modes are used in the final analysis.) The spectral values do not include the response modification coefficient R. This term is brought into the analysis later. Also, the system damping, assumed to be 5 percent critical, is included in the development of the response spectrum, and it need not be considered elsewhere in this analysis.

Table G20–2 Modal Properties for Six-Story Structure

1	2	3	4	5	6	7
Mode	w (rad/s)	T (s)	G	m (kips-s²/in.)	Accumulated Effective Modal Mass (kips-s²/in.)	Accumulated Mass/Total Mass (% of Total)
1	3.36	1.871	1.669	11.179	11.179	71.1
2	7.14	0.880	−0.957	2.754	13.933	88.6
3	12.0	0.524	−0.382	1.231	15.164	96.4
4	20.6	0.305	0.276	0.242	15.406	98.7
5	26.4	0.238	−0.188	0.203	15.609	99.3
6	32.5	0.194	0.110	0.114	15.723	100.0

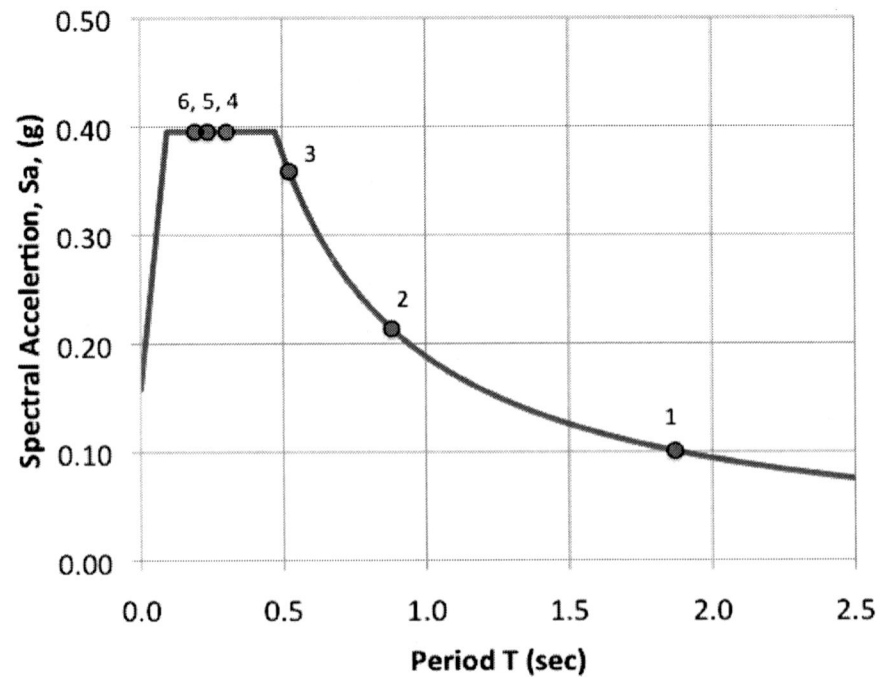

Figure G20–4 Elastic Response Spectrum.

Table G20-3 Spectral Accelerations and Displacements

1	2	3	4	5	6
Mode	Period T (s)	S_a (g)	S_a (in./s²)	S_d (in.)	ΓS_d (in.)
1	1.871	0.100	38.83	3.44	5.74
2	0.880	0.214	82.58	1.62	−1.55
3	0.524	0.359	138.69	0.94	−0.36
4	0.305	0.395	152.6	0.360	0.100
5	0.238	0.395	152.6	0.219	−0.041
6	0.194	0.395	152.6	0.144	0.016

The spectral displacements are determined from the spectral accelerations through the use of the formula

$$S_{di} = \frac{S_{ai}}{\omega_i^2}g = \frac{S_{ai}T_i^2}{4\pi^2}g \qquad \text{(G20-4)}$$

where g is the acceleration of gravity = 386.4 in./s². The spectral displacements at the system's first three modal periods are provided in column 5 of **Table G20–3**.

Column 6 of **Table G20–3** provides the product of the modal participation factor and the spectral displacement for each mode. Because of the scaling of the mode shapes to produce a maximum modal displacement of 1.0 in each mode, the terms in column 6 are a direct indicator of each mode's contribution to the displacements at each floor level of the structure. As expected, the first mode is responsible for the vast majority of displacement in the system.

20.4 Computation of Story Displacements and Story Drift

The next step in the analysis is to determine the displacements in each mode. These displacements are computed as follows:

$$\delta_i = \Gamma_i S_{di} \phi_i \qquad \text{(G20-5)}$$

The displacements are then combined using the square root of the sum of the squares (SRSS) to determine the total displacements at each story. However, story drifts should not be determined from the SRSS of the story displacements. Instead, the drifts should be determined for each mode, and then these story drifts are combined using SRSS. The calculations for story displacement and story drift are provided in **Table G20–4**.

The displacements and story drifts in **Table G20–4** are the elastic values and have not been modified to account for the expected inelastic behavior. To modify for inelastic effects, the values must be multiplied by the quantity (C_d/R). The modified story drifts are provided in **Table G20–5** together with the limiting values of story drift that are given in Table 12.2-1 of ASCE 7. The story drift limit for this Occupancy Category II building is 0.02 times the story height. It appears that the story drifts are well below the drift limit, particularly in the lower levels.

The displacements in **Tables G20–4** and **G20–5** are based on the computed periods of vibration from the eigenvalue analysis, and not on the empirical period T_a or $C_u T_a$. Indeed, these empirical periods have not yet been computed. Section 12.9.4 allows the drift calculations to be based on the computed period without scaling, which might be required for member forces. This result is consistent with Section 12.8.6.2, which allows the displacements computed by the ELF method to be based on the computed period.

20.5 Story Forces and Story Shears

The elastic modal story forces are determined from the following equation, in which K is the 6-DOF's stiffness matrix of the system, and δ_i is the modal story displacement vector for mode i:

$$F_i = K\delta_i \tag{G20-6}$$

The elastic story shears are determined for each mode from the elastic story forces, and the total elastic story shears are then computed from the SRSS of the elastic modal story shears. To account for inelastic behavior and for importance, the elastic story shears must be modified by multiplying each value by the ratio (I/R). The elastic story forces are shown in **Table G20–6(a)** and the elastic and inelastic story shears are provided in **Table G20–6(b)**. The inelastic shears are shown in the next to the last column of **Table G20–6(b)**. The design story shears, shown in the last column, are based on the scaling requirements of Section 12.9.4 of ASCE 7. The scaling procedure is described in the following text.

The inelastic base shear, which is the same as the first-story inelastic shear (115.3 kips), is computed on the basis of spectral ordinates that are in turn based on the computed periods of vibration for the system. Section 12.9.4 states that the design base shear must not be less than 85 percent of the base shear computed using the empirical period of vibration, $C_u T_a$.

For a moment resisting frame, T_a is computed according to Eq. 12.8-7:

$$T_a = C_t h_n^x \tag{12.8-7}$$

Table G20–4 Elastic Displacements and Story Drifts

Story	Elastic Story Displacements (in.)				Elastic Interstory Drifts (in.)			
	Mode 1	Mode 2	Mode 3	SRSS	Mode 1	Mode 2	Mode 3	SRSS
6	5.745	−1.548	0.349	5.960	1.126	−1.187	0.590	1.739
5	4.619	−0.362	−0.241	4.639	1.083	−0.827	0.127	1.368
4	3.536	0.465	−0.368	3.585	0.987	−0.231	−0.347	1.071
3	2.549	0.696	−0.021	2.643	1.034	0.060	−0.315	1.083
2	1.515	0.635	0.294	1.669	0.762	0.272	0.059	0.811
1	0.753	0.364	0.236	0.869	0.753	0.364	0.236	0.869

Table G20–5 Inelastic Story Drifts and Drift Limits

1	2	3	4	5
Story	SRSS Drift (in.)	SRSS Drift × (C_d/R) (in.)	Drift Limit (in.)	Ratio (3/4)
6	1.739	1.546	3.60	0.429
5	1.368	1.216	3.00	0.405
4	1.071	0.952	3.00	0.317
3	1.083	0.963	3.00	0.321
2	0.811	0.721	3.00	0.240
1	0.869	0.773	3.60	0.215

Note: $C_d = 4.0$ and $R = 4.5$.

Table G20–6a Elastic Story Forces

Story	Elastic Story Forces (kips)		
	Mode 1	Mode 2	Mode 3
6	83.9	−102.2	65.0
5	70.8	−25.1	−47.1
4	103.2	61.4	−137.2
3	76.3	94.2	−8.1
2	66.4	125.8	164.3
1	33.5	73.7	133.8

Table G20–6b Story Shears

Story	Elastic Story Shears (kips)				Inelastic Story Shear (kips)	Design Story Shears (kips)
	Mode 1	Mode 2	Mode 3	SRSS		
6	83.9	−102.2	65	147.3	32.9	42.8
5	154.7	−127.3	17.9	201.1	44.7	58.1
4	257.9	−65.9	−119.2	291.7	66.0	85.8
3	334.2	28.3	−127.3	358.8	79.8	103.7
2	400.6	154.1	36.9	430.8	95.5	124.1
1	434.1	227.4	170.7	518.9	115.3	149.9

Using the coefficients for a steel moment frame from Table 12.8-2 and a height of 80 ft,

$$T_a = 0.028(80)^{0.8} = 0.93 \text{ s}$$

Interpolating from Table 12.8-1 with $S_{D1} = 0.188\ g$, $C_u = 1.52$, the upper limit on period is

$$C_u T_a = 1.52 \times 0.93 = 1.41 \text{ s}$$

This period is somewhat less than the computed first mode period of vibration, which is 1.87 s. Using Eq. 12.8-1 with $W = 6{,}075$ kips (for half of the building) and Eq. 12.8-3,

$$C_s = \frac{S_{D1}}{T\left(\dfrac{R}{I}\right)} = \frac{0.188}{1.41\left(\dfrac{4.5}{1.0}\right)} = 0.029$$

Checking Eq. 12.8-5, C_s shall not be less than $0.044\ S_{DS} = 0.044(0.395) = 0.017$, so Eq. 12.8-3 governs, giving

$$V = C_s W = 0.029 \times 6{,}075 = 176 \text{ kips}$$

According to Section 12.9.4, the design base shear must not be less than 85 percent of the ELF base shear. For the current example, this is $0.85(176) = 149.6$ kips. This value is greater than the inelastic base shear of 115.3 kips, so the inelastic shears must be scaled by the ratio $149.6/115.3 = 1.30$ to obtain the design shears. The design story shears are shown in the last column of **Table G20–6(b)**.

20.6 Computation of Design Member Forces

Elastic member forces are obtained by computing the member forces in each mode and then taking the SRSS of these values, on an element-by-element basis. The elastic member forces are obtained by loading the full 66-DOF system with the elastic modal story forces shown in **Table G20–6(a)**. Design forces are then obtained by multiplying the elastic forces by ($I/R = 1.0/4.5$), and then (for the current problem) by the scale factor of 1.3.

This procedure is illustrated by use of **Figs. G20–5(a)** through **(c)**. Each figure shows the modal loading and the resulting elastic member forces.

For the beam, the combined shears and moments are determined from SRSS as follows:

$$\text{Elastic moment at left end} = \sqrt{5078^2 + 1122^2 + 638^2} = 5239 \text{ in.-kips}$$
$$\text{Elastic moment at right end} = \sqrt{5064^2 + 1122^2 + 632^2} = 5225 \text{ in.-kips}$$
$$\text{Elastic shear} = \sqrt{28.2^2 + 6.23^2 + 3.53^2} = 29.1 \text{ kips}$$
$$\text{Design moment at left end} = 1.3(5{,}239)(1.0)/(4.5) = 1{,}513 \text{ in.-kips}$$
$$\text{Design moment at right end} = 1.3(5{,}225)(1.0)/(4.5) = 1{,}509 \text{ in.-kips}$$
$$\text{Design shear} = 1.3(29.1)(1.0)/(4.5) = 8.41 \text{ kips}$$

The values for the column are

$$\text{Elastic moment at bottom} = \sqrt{5858^2 + 37.6^2 + 2500^2} = 6369 \text{ in.-kips}$$
$$\text{Elastic moment at top} = \sqrt{5214^2 + 698^2 + 1626^2} = 5506 \text{ in.-kips}$$
$$\text{Elastic shear} = \sqrt{73.8^2 + 4.4^2 + 27.5^2} = 78.9 \text{ kips}$$
$$\text{Design moment at bottom} = 1.3(6{,}369)(1.0)/(4.5) = 1{,}840 \text{ in.-kips}$$
$$\text{Design moment at top} = 1.3(5{,}506)(1.0)/(4.5) = 1{,}591 \text{ in.-kips}$$
$$\text{Design shear} = 1.3(78.9)(1.0)/(4.5) = 22.8 \text{ kips}$$

20.7 Equivalent Lateral Force Analysis

The same frame that was analyzed by the MRS method was reanalyzed using the ELF method. The results of the analysis are presented in **Tables G20–7** and **G20–8**. Only minimal details are provided for the ELF analysis because the purpose of showing the ELF results is for comparison with the MRS results. A detailed example that considers only the ELF procedure is provided in Example 18.

Analysis for story forces and story shears is presented in **Table G20–7**. These forces are based on the upper limit empirical period of vibration $T = C_u T_a = 1.41$ s. As described in the previous section, the design base shear is 176 kips. Distribution of forces along the height is based on Eqs. 12.8-11

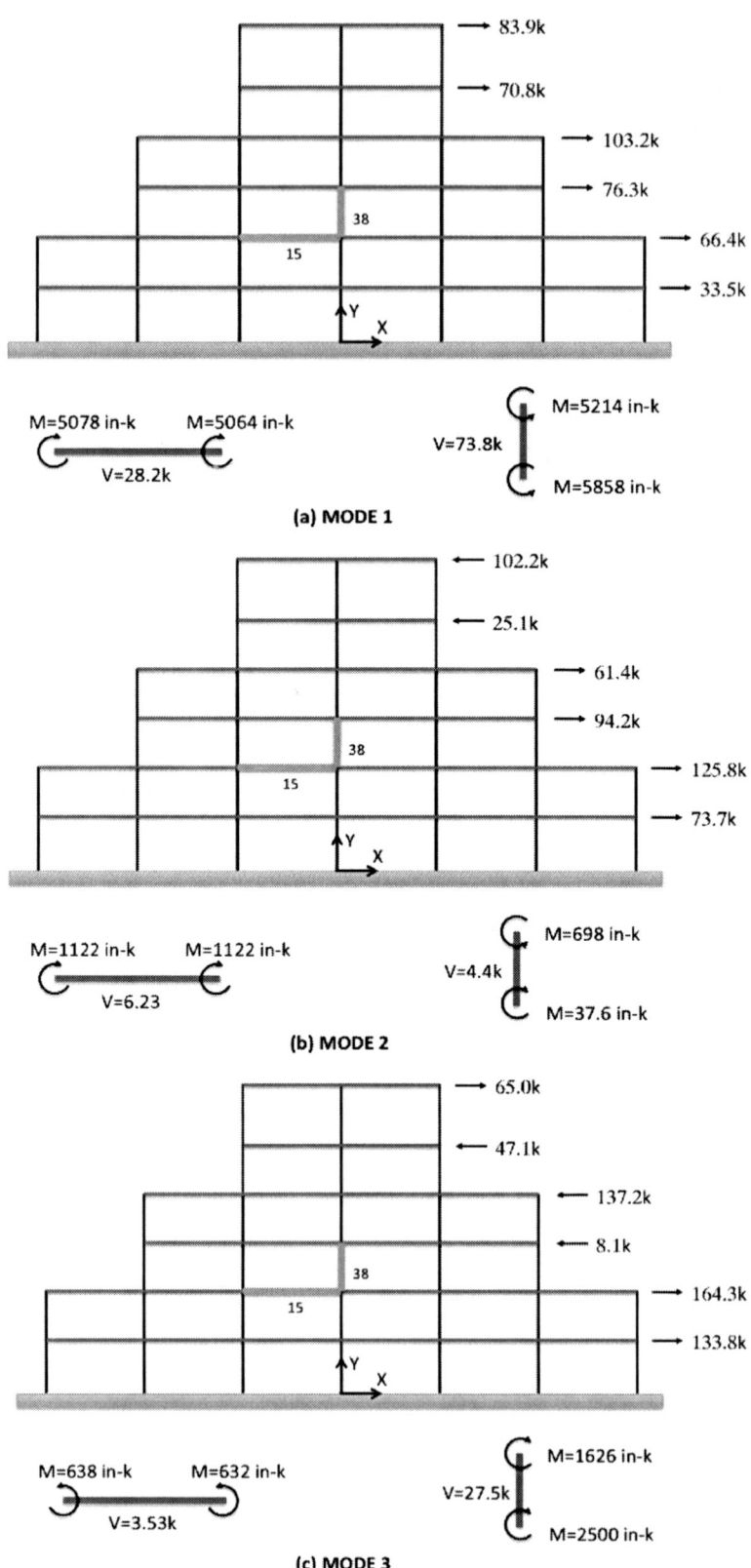

Figure G20–5 Computing Member Forces.

Table G20-7 ELF Analysis: Story Shears

1	2	3	4	5	6	7	8
Story	H	h (ft)	W (kips)	Wh^k	Wh^k/Total	F (kips)	V (kips)
6	15	80.0	550	465,361	0.234	41.1	41.1
5	12.5	65.0	525	353,431	0.177	31.2	72.3
4	12.5	52.5	1,000	482,446	0.242	42.6	114.9
3	12.5	40.0	1,025	323,547	0.162	28.6	143.5
2	12.5	27.5	1,500	263,907	0.132	23.3	166.8
1	15	15.0	1,525	104,225	0.052	9.2	176.0
Total	80	—	6,125	1,992,917	1.000	176.0	—

Table G20-8 ELF Analysis: Story Drifts

1	2	3	4	5	6	7	8
Story	H	h (ft)	W (kips)	Wh^k	Wh^k/Total	F (kips)	$C_d\Delta$ (in.)
6	15	80.0	550	2,303,662	0.292	39.6	2.03
5	12.5	65.0	525	1,481,108	0.188	25.4	1.81
4	12.5	52.5	1,000	1,878,855	0.238	32.3	1.48
3	12.5	40.0	1,025	1,147,763	0.145	19.7	1.46
2	12.5	27.5	1,500	823,192	0.104	14.1	1.00
1	15	15.0	1,525	264,037	0.033	4.53	0.952
Total	80	—	6,075	7,898,596	1.000	135.6	—

and 12.8-12 with the exponent $k = 1.56$ for the period of vibration of 1.41 s. Column 8 of **Table G20-7** contains the design story shears.

Member forces were computed for the structure loaded with the forces in column 7 of **Table G20-7**. The values obtained for the beam and column indicated in **Fig. G20-5** are as follows:

For the Beam

    Design moment at left end = −2,148 in.-kips (negative is clockwise)
    Design moment at right end = −2,142 in.-kips
    Design shear = 11.9 kips

For the Column

> Design moment at bottom = 2,539 in.-kips
> Design moment at top: = 2,227 in.-kips
> Design shear: =31.8 kips

Section 12.8.6.2 states that story drifts may be based on the computed period of vibration. For the system under consideration, the fundamental period from the eigenvalue analysis is 1.871 s, and the computed base shear using this period is 135.6 kips. For computing displacements, lateral forces are obtained using Eqs. 12.8-11 and 12.8-12 with the exponent k based on the computed period. In this case, $k = 1.903$, and the resulting story forces are shown in column 7 of **Table G20–8.** The interstory drifts resulting from the application of these forces to the structure are shown in column 8 of **Table G20–8.** These drifts include the deflection amplification factor $C_d = 4.0$.

The MRS and ELF results are compared in **Tables G20–9(a)** through **G20–9(c)**. In general, the MRS method produces shears, drifts, and member forces in the neighborhood of 70 percent of the values obtained using ELF. This difference indicates that for this structure, the use of MRS provides substantial economy, compared to ELF.

20.8 Three-Dimensional Modal Response Spectrum Analysis

It is beyond the scope of this guide to present a detailed example of a three-dimensional modal response spectrum analysis. However, certain aspects of such an analysis are pertinent, and these aspects are excerpted from an example developed by the author for the *NEHRP Recommended Provisions Design Examples* document, available on compact disc from FEMA (FEMA 2006).

The building under consideration is a 12-story steel building with one basement level. The structural system is a perimeter moment resisting space frame. A three-dimensional wire-frame drawing of the building, as modeled with the SAP2000 program (CSI 2009) is shown in **Fig. G20–6**. Although it is not clear from the drawing, the basement level of the building was explicitly modeled, and thick-shell elements were used to represent the basement walls. Floor diaphragms were modeled as rigid in-plane and flexible out-of-plane. For analysis, the structure was assumed to be fixed at the bottom of the basement level. (For computing the approximate period T_a (Eq. 12.8-7), the height of the building would be measured from the top of the basement walls.)

The properties of the first 10 mode shapes are shown in **Table G20–10**. For each mode, the periods of vibration are given, as well as modal direction factors. The modal direction factors indicate the predominant direction of the mode. The first mode, with a period of 2.867 s, is a translational response in the X direction. The second mode is translational in the Y direction, and the third mode is torsion. The fourth and fifth modes are predominantly lateral, but the sixth and higher modes have significant lateral–torsional coupling. The first eight mode shapes are plotted in **Fig. G20–7**.

Table G20-9a Comparison of MRS and ELF Design Story Shears

1	2	3	4
Story	MRS Shear (kips)	ELF Shear (kips)	MRS/ELF Shear
6	42.8	41.1	1.041
5	58.1	72.3	0.804
4	85.8	114.9	0.747
3	103.7	143.5	0.724
2	124.1	166.8	0.744
1	149.9	176.0	0.85

Table G20-9b Comparison of MRS and ELF Story Drifts

1	2	3	4
Story	MRS Drift (in.)	ELF Drift (in.)	MRS/ELF Drift
6	1.546	2.03	0.759
5	1.216	1.81	0.670
4	0.952	1.48	0.641
3	0.963	1.46	0.661
2	0.721	1.00	0.718
1	0.773	0.952	0.811

Table G20-9c Comparison of MRS and ELF Member Design Forces

1	2	3	4
Item	MRS	ELF	MRS/ELF
Beam moment left (in.-kips)	1,513	2,148	0.704
Beam moment right (in.-kips)	1,509	2,142	0.704
Beam shear (kips)	8.41	11.9	0.705
Column moment bottom (in.-kips)	1,840	2,539	0.724
Column moment top (in.-kips)	1,591	2,227	0.714
Column shear (kips)	22.8	31.8	0.717

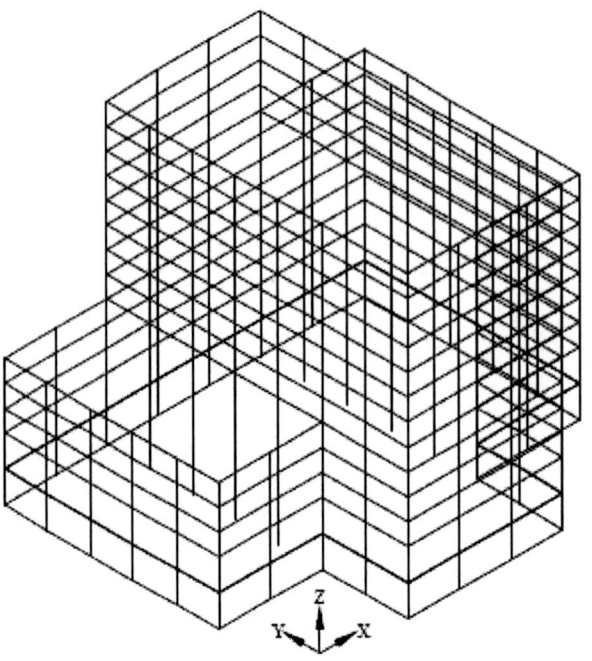

Figure G20–6 Wire-Frame Model of 12-Story Building (Lower Level Is the Basement Wall).

The effective modal masses for the first 10 modes are provided in **Table G20–11**. For each mode, the effective mass for that mode in the direction of interest and the accumulated mass in that direction are given. For the third mode, for example, the effective mass in the X direction is 0.34 percent of the total mass in the X direction, and the accumulated X direction mass in modes 1, 2, and 3 is 64.9 percent of the total X-direction mass.

Section 12.9.1 of ASCE 7 requires that enough modes be included to capture at least 90 percent of the total mass in each direction. Clearly, this requirement is not achieved with 10 modes. In fact, it would take 40 modes to satisfy this requirement, 40 is more modes that one would expect for a 12-story building with rigid diaphragms (10 modes would usually be sufficient). The reason that 10 modes are not sufficient is based on the fact that almost 18 percent of the total mass of the structure is represented by the subgrade level of the building, including the basement walls and the grade-level diaphragm. In this sense, the spirit of ASCE 7 could be met with only 10 modes because these modes capture almost 100 percent of the dynamically excitable mass in the above-grade portion of the structure. If it is desired to determine the stresses and forces in the basement walls, at least 40 modes would be required.

Also, the complete quadratic combination method of modal combination should be used for this structure because the midlevel modes have a high degree of lateral–torsional coupling. Modal combination requirements are provided in Section 12.9.3 of ASCE 7.

Table G20–10 Modal Properties for 12-Story Steel Moment Frame Building

Mode	Period (s)	Modal Direction Factor		
		X	Y	Torsion
1	2.867	99.2	0.7	0.1
2	2.745	0.8	99.0	0.2
3	1.565	1.7	9.6	88.7
4	1.149	98.2	0.8	1.0
5	1.074	0.4	92.1	7.5
6	0.724	7.9	44.4	47.7
7	0.697	91.7	5.23	3.12
8	0.631	0.3	50.0	49.7
9	0.434	30.0	5.7	64.3
10	0.427	70.3	2.0	27.7

Table G20–11 Effective Modal Masses for 12-Story Steel Moment Frame Building

Mode	Effective Modal Masses (% of Total)					
	X (Mode)	X (Accumulated)	Y (Mode)	Y (Accumulated)	T (Mode)	T (Accumulated)
1	64.04	64.0	0.46	0.5	0.04	0.0
2	0.51	64.6	64.25	64.7	0.02	0.1
3	0.34	64.9	0.93	65.5	51.06	51.1
4	10.78	75.7	0.07	65.7	0.46	51.6
5	0.04	75.7	10.64	76.3	5.30	56.9
6	0.23	75.9	1.08	77.4	2.96	59.8
7	2.94	78.9	0.15	77.6	0.03	59.9
8	0.01	78.9	1.43	79.0	8.93	68.8
9	0.38	79.3	0.00	79.0	3.32	71.1
10	1.37	80.6	0.01	79.0	1.15	72.3

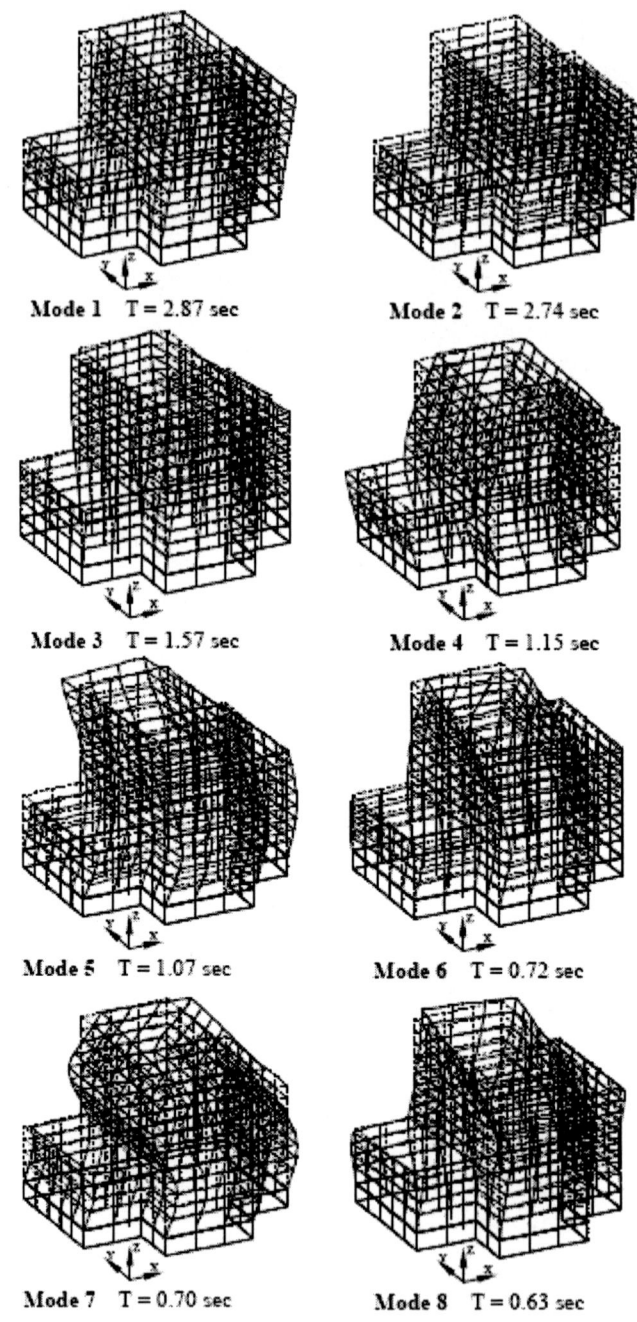

Figure G20-7 First Eight Mode Shapes of a 12-Story Steel Building.

Example 21
Diaphragm Forces

This example discusses Section 12.10 of ASCE 7, which covers the computation of in-plane floor diaphragm forces, including collector and chord elements. Diaphragm forces caused by both inertial and system effects are included.

The structure considered in this example is a six-story office building located near Memphis, Tennessee. The pertinent information for the building and the building site are as follows:

Site Class = B
$S_S = 0.6\ g$
$S_1 = 0.2\ g$
$F_a = 1.0$ (from Table 11.4-1)
$F_v = 1.0$ (from Table 11.4-2)
$S_{DS} = (2/3)\ S_S \times F_a = 0.400\ g$ (11.4-1 and 11.4-3)
$S_{D1} = (2/3)\ S_1 \times F_v = 0.133\ g$ (11.4-2 and 11.4-4)
Occupancy Category = II (from Table 1-1)
Importance Factor $I = 1.0$ (from Table 11.5-1)
Seismic Design Category = C (from Tables 11.6-1 and 11.6-2)

A plan view of the structural system for the building is shown in **Fig. G21-1**. Intermediate steel moment frames resist all of the forces in the east–west direction, and a dual intermediate moment frame–special concentrically braced frame system is used in the north–south direction. The lowest story has a height of 15 ft, and the upper stories each have a height of 12.5 ft. The total seismic weight of the system W is 7,500 kips.

This example considers loads acting in the north–south direction only. The design values for the dual system are determined from Table 12.2-1 and are summarized as follows:

$R = 6$
$\Omega_o = 2.5$
$C_d = 5.0$
Height limit = None

The period of vibration from structural analysis is 0.73 s. This period controls over the upper limit period $T = C_u T_a$, which is 0.85 s for this structure.

The seismic response coefficient is taken as the larger of the values computed from Eqs. 12.8-3 and 12.8-5:

$$C_s = \frac{S_{D1}}{T(R/I)} = \frac{0.133}{0.73(6/1)} = 0.0304$$

$$C_s = 0.044 S_{DS} I = 0.044(0.40)(1) = 0.0176 > 0.01$$

The design base shear (Eq. 12.8-1) is

$$V = C_s W = 0.0304(7,500) = 228 \text{ kips}$$

The floor deck consists of a 4.5-in. concrete slab over metal deck. This slab must resist both inertial forces and forces developed because of shear transfer between the lateral force resisting elements. Also considered in the analysis are the collector elements (drag struts) and diaphragm chords. These elements are shown in **Fig. G21-1**. The elements shown as diaphragm chord elements act as collector elements when seismic forces act in the east–west direction.

The inertial forces at a given level x are computed in accordance with Section 12.10.1.1 and Eq. 12.10-1:

$$F_{px} = \frac{\sum_{i=x}^{n} F_i}{\sum_{i=x}^{n} w_i} w_{px} = q_{px} w_{px} \qquad (12.10\text{-}1)$$

where F_i is the lateral force applied to level i, w_i is the weight at level i, w_{px} is the weight of the diaphragm (or portion thereof) at level x, and n is the number of levels. The term q_{px} is not explicitly used in ASCE 7, but it is a convenient

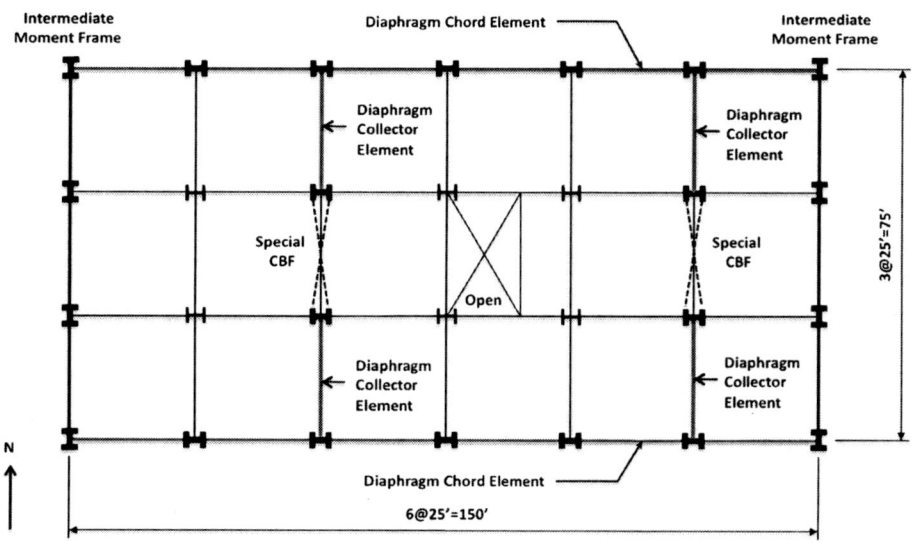

Figure G21–1 Typical Floor Plan of a Six-Story Building.

parameter to determine. The force F_{px} that is obtained from Eq. 12.10-1 shall not be less than $0.2S_{DS}Iw_{px}$ but need not exceed $0.4S_{DS}Iw_{px}$. The results of the analysis for the coefficient θ is given in **Table G21–1**. The minimum value given by $0.2S_{DS}I$ controls at each level.

On the basis of the above calculations, each level of the diaphragm must be designed for a force of 0.08 times the seismic weight at the level of interest. For the fifth level, for example, the total diaphragm force is $0.08(1,250) = 100.0$ kips. From a structural analysis perspective, the most accurate way to determine the stresses and forces in the various components of the diaphragm would be to model the diaphragm with shell elements and distribute the 100-kips force to the individual nodes of the element on a "tributary mass" basis. Thus, the diaphragm would be modeled as semirigid, even though Section 12.3.1.2 would define this diaphragm as rigid. Other analysis approaches are also available, including the linear collector method, the distributed collector method, and the strut and tie modeling method. See Sabelli et al. (2009) for a description of these methods.

Additional comments regarding the analysis and design of the diaphragm components are as follows:

1. Discontinuities in the lateral load resisting elements above and below the diaphragm cause forces to be transferred between the lateral load resisting elements and impose in-plane forces in the diaphragm. These forces must be added to the forces developed from the diaphragm forces F_{px} that have been determined with Eq. 12.10-1. In accordance with Section 12.12.1.1, elements resisting the transfer forces must be designed with the redundancy factor ρ that has been determined for

Table G21–1 Diaphragm Force Coefficients for a Six-Story Building

1	2	3	4	5	6	7	8	9
Level	w (kips)	F (kips)	Σw (kips)	ΣF (kips)	$\theta = \Sigma F / \Sigma w$	$q_{min} = 0.2S_{DS}I$	$q_{max} = 0.4S_{DS}I$	Controlling F_{px} (kips)
6	1,150	62.5	1,150	62.5	0.054	0.080	0.160	92.0
5	1,250	55.8	2,400	118.3	0.049	0.080	0.160	100.0
4	1,250	44.0	3,650	162.3	0.044	0.080	0.160	100.0
3	1,250	32.5	4,900	194.8	0.040	0.080	0.160	100.0
2	1,250	21.4	6,150	216.2	0.035	0.080	0.160	100.0
1	1,350	11.8	7,500	228.0	0.030	0.080	0.160	108.0

the structural system, but the forces caused by the application of F_{px} alone may be designed with a redundancy factor of 1.0.

2. In SDC C and above, collector elements must be designed using the overstrength factor Ω_o that has been assigned to the structural system resisting forces in the direction of F_{px}.
3. Special care has to be taken to produce realistic diaphragm, collector, and chord forces when the diaphragm is modeled as rigid in a three-dimensional structural analysis.
4. Recovering diaphragm forces from a modal response spectrum (MRS) analysis is not straightforward. It is reasonable to use the approach outlined in Section 12.10 to determine diaphragm forces, even when MRS has been used for the analysis of the lateral load resisting system. However, as mentioned in point 1 above, the total diaphragm forces must include the transfer forces, if present, and these forces must be recovered from the MRS analysis.

Frequently Asked Questions

The following questions were provided by practicing engineers who use ASCE 7 on a regular basis. Some of the issues in these questions are addressed in the main body of the guide, and others are not. With few exceptions, the answers provided below are made without reference to the material in the guide, thereby allowing this FAQ section be a stand-alone reference.

1. **Does the interconnection requirement of Section 12.1.3 apply across any section cut a designer might draw across a diaphragm?**

Section 12.1.3 provides requirements for a continuous load path and interconnection of all parts of the structure. These requirements would apply across any section cut through a diaphragm. Additionally, the diaphragm must be analyzed and designed in accordance with Section 12.10.

2. How are story stiffnesses calculated when determining if a vertical structural irregularity of Type 1a or 1b exists?

Table 12.3-2 discusses the conditions under which a vertical structural irregularity exists. Stiffness irregularities (Types 1a and 1b) occur when the lateral stiffness of one story is less than a certain percentage of the lateral stiffness of the story above it (or less than a certain percentage of the average stiffness of the three stories above it). Note, however, that it is only necessary to determine if such irregularities exist for buildings in SDC D and above because there are no consequences of stiffness irregularities in buildings assigned to SDC C and lower.

To address the issue of determining story stiffness, consider the structure shown in **Fig. FAQ–1**. This structure is a one-bay moment resisting frame. The column stiffness is the same at each level and the beam stiffness is the same at each level. Three configurations were analyzed, with each system having a different beam-to-column stiffness ratio as follows:

beam-to-column stiffness ratio = 0.01:   Structure behaves like a vertical cantilever.
beam-to-column stiffness ratio = 1.00:   Structure behaves like a moment frame.
beam-to-column stiffness ratio = 100:   Structure behaves like a shear frame.

For all systems, the columns are fixed at the base. Two methods were used to determine the lateral stiffness of the story:

Method 1. Load the structure with a unit lateral force at each level (all levels loaded simultaneously), compute the interstory drift at each level, and determine the stiffness at each story as the story shear divided by the story drift. Only one loading is required for this method.

Method 2. Load the structure with a positive unit force at level i and a negative unit force at level $i - 1$, compute the interstory drift between the two loaded levels, and invert to obtain the stiffness of the story between levels i and $i - 1$. Ten separate loadings are required for this method.

The results of the analysis are shown in **Fig. FAQ–2**. Part (a) of this figure is for the system with a beam-to-column stiffness ratio of 0.01. The two methods produce dramatically different story stiffness and for each method, the stiffness increases significantly at the lowest level. This statement is particularly true when Method 2 is used. This increase in stiffness at the lowest level occurs because of the fixed support at the base. The change in stiffness along the height is caused by the support condition and by rigid body rotation in the upper floors, and not by any actual variation in the stiffness of the structural system.

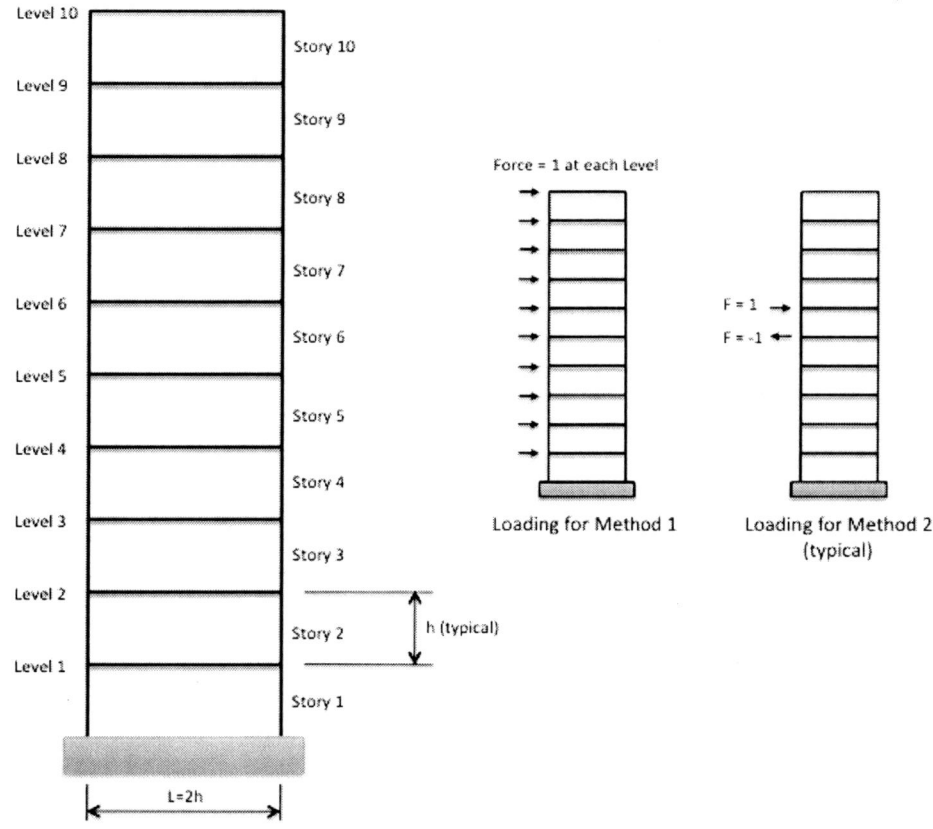

Figure FAQ-1 System and Loading for Determination of Story Stiffness.

The results for the system with the beam-to-column stiffness ratio of 1.0 is presented in **Fig. FAQ-2(b)**. As before, the results are dramatically different for the two methods of analysis, with Method 2 providing the larger story stiffness. For Method 1, there is a large variation in stiffness along the height, but for Method 2 there is a relatively uniform stiffness, except for the first level. The increase in stiffness at the first level is caused by the fixed base condition.

The results for the analysis of the system with relatively stiff beams are shown in **Fig. FAQ-2(c)**. Here the two methods give similar results, with Method 2 producing a consistently greater stiffness than Method 1. Although there is some influence of the fixed base condition on the results, this difference is less significant that it was for the other two beam-to-column stiffness ratios.

Method 2 appears to be producing better results than Method 1 because the story stiffnesses reported by Method 2 are more uniform (as would be expected for a structure with uniform properties along the height). The results of Method 1 are highly influenced by the accumulated rotations of the stories below the story of interest. For example, for the system with the

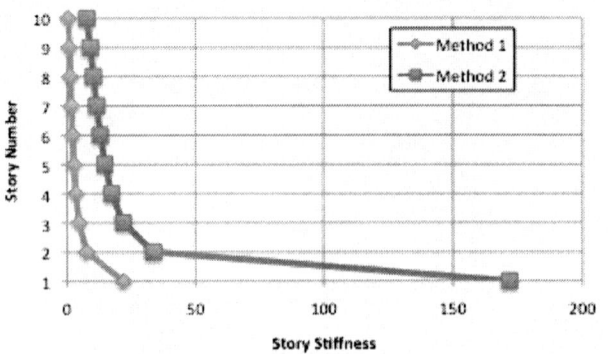

(a) Beam-to-Column Stiffness Ratio = 0.01.

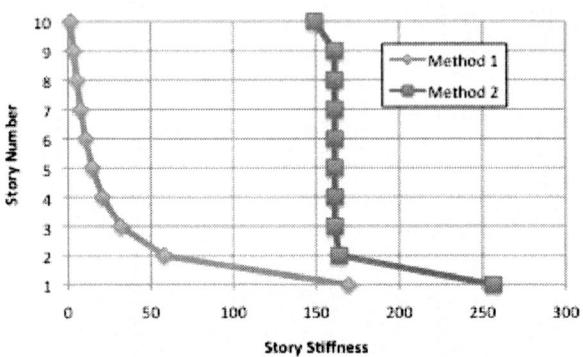

(b) Beam-to-Column Stiffness Ratio = 1.0.

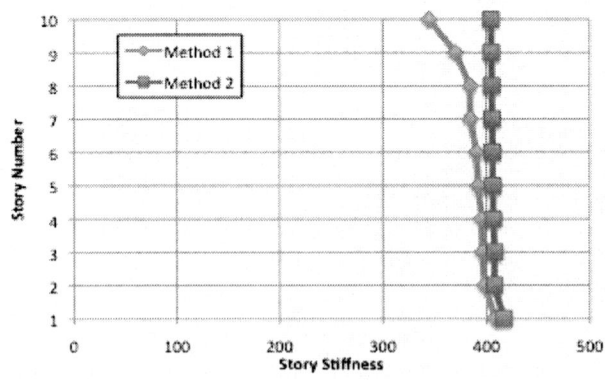

(c) Beam-to-Column Stiffness Ratio = 100.

Figure FAQ-2 Results of Stiffness Analysis of a 10-Story Building System.

lowest beam-to-column stiffness ratio, almost all of the drift in the upper levels is caused by rigid body rotation.

The basic conclusions from the analysis are as follows:

1. Method 1 should not be used to determine story stiffness.
2. Method 2 is preferred, but the computed stiffnesses at the lower levels may be artificially high because of the fixed boundary conditions, and the stiffness at the upper levels may be artificially low because of the presence of rigid body rotations. The method appears to be reliable for moment resisting frames, but it may produce unrealistic estimates of story stiffness in systems that deform like a cantilever (e.g., tall, slender shear walls and braced frames).

3. How are story strengths calculated when determining if a vertical structural irregularity of Type 5a or 5b exists (vertical structural irregularities of types 5a and 5b in Table 12.3-2)?

It is difficult if not impossible to calculate the strength of a story of a lateral load resisting seismic system. The computed strength depends on the loading pattern, the location of yielding throughout the story, and on the capacities of the elements that are yielding. The element capacities are a function of the materials used and the details of the cross section, and of the forces that act on the section. For example, the flexural strength of a reinforced concrete shear wall is a function of the axial compressive force in the wall. Similarly, the flexural capacity of a steel or concrete column is a function of the axial force in the column. Shear capacities of concrete sections are also a function of the axial force in the section.

It is possible to estimate the story capacities of some simple systems, such as those shown in **Fig. FAQ–3**. For system (a), which is a braced frame system, the story capacity can be based on the strength of the braces and would be as follows:

$$V_u = (F_{uC} + F_{uT}) \cos \phi \tag{F1}$$

where F_{uC} and F_{uT} are the compressive and tensile capacities of the braces, respectively, and ϕ is the angle shown in the figure. This capacity assumes that the columns do not yield axially and that they have a moment release (moment-free hinge) at the top and bottom of the story. If it is assumed that the columns also yield, the strength from a column mechanism (**Fig. FAQ–3(b)** and Eq. F2 below) may be added to the strength obtained from Eq. F1.

For a moment resisting system, such as shown in **Fig. FAQ–3(b)**, the story capacity may be based on a sway mechanism. This method is based on the assumption that plastic hinges form in the top and bottom of each

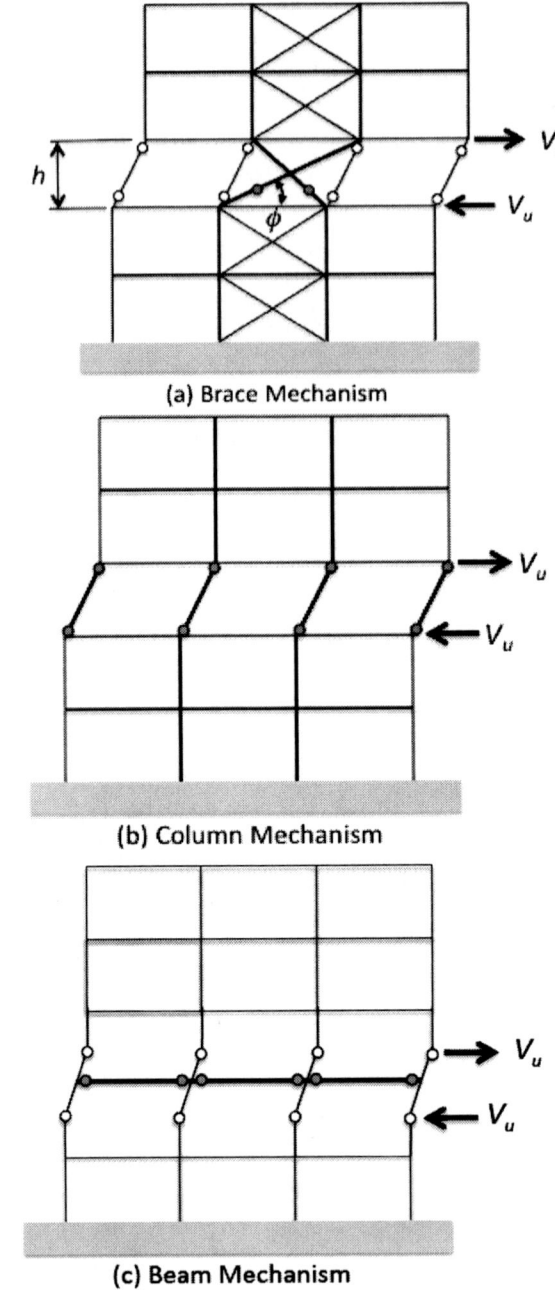

Figure FAQ-3 Three Mechanisms for Computing Story Strength.

column of a particular story. If the flexural capacities of the columns are known, the story strength may be obtained as

$$V_u = \frac{2}{h} \sum_{i=1}^{ncols} M_{uC,i} \tag{F2}$$

where h is the story height, $ncols$ is the number of columns in the story, and M_{uC} is the flexural capacity of the column hinges at the top and bottom of the columns (which may be a function of the axial force in the column). This type of mechanism might form in columns of ordinary and intermediate moment frames, but it is unlikely to occur in special moment frames because of strong column–weak beam design requirements.

A second type of story capacity may be computed on the basis of the beam strengths. The mechanism for computing the story capacity is shown in **Fig. FAQ–3(c)**. This computed capacity is

$$V_u = \frac{1}{h} \sum_{i=1}^{nbays} [M_{uB,i}^+ + M_{uB,i}^-] \tag{F3}$$

where $nbays$ is the number of bays, M_{uB}^+ is the positive moment flexural capacity at one end of the beam, M_{uB}^- is the negative moment capacity at the other end of the beam, and $nbays$ is the number of bays. However, a mechanism consisting of plastic hinges at each end of each beam in a single story is impossible (without loss of continuity in the columns above and below the level in question). Equations similar to Eqs. F2 and F3 are discussed in Part C3 of the commentary in *Seismic Provisions for Structural Steel Buildings* (AISC 2005b).

There is no straightforward way to compute the story shear capacity of a shear wall. Determination of the story capacity of combined systems and dual systems is also problematic.

ASCE 7 has only two consequences when weak-story irregularities occur. The first of these consequences is given in Section 12.3.3.1, which prohibits structures in SDC E and F from having a Type 5a or 5b vertical irregularity and structures in SDC D from having a Type 5b irregularity. The second consequence is given by Section 12.3.3.2, which states that buildings with vertical irregularity Type 5b must be limited in height to 30 ft (with certain exceptions). Also, weak story irregularities do not prohibit one from using the equivalent lateral force method of analysis, whereas soft story irregularities in SDC D and above buildings may prohibit the use of ELF.

Aside from determining if weak story irregularity exists, story shear capacity may also be needed in association with computing the redundancy factor (Section 12.3.4) and for determining the limiting value of the stability coefficient (Eq. 12.8-17).

4. Why are forces for columns supporting discontinuous braced frames amplified by the overstrength coefficient Ω_o, but not columns in braced frames that are continuous?

This requirement, from Section 12.3.3.3, applies only to structures with an in-plane (Type 4 horizontal irregularity) or out-of-plane offset (Type 4 vertical irregularity) in the lateral load resisting system. Experience from previous earthquakes has indicated that such irregularities impose extreme demands on the portion of the structure below the irregularity, and such irregularities have been identified as a significant contributor to the partial or complete collapse of structures during earthquakes. The amplification factor serves as a penalty, discouraging the use of such irregularities and reducing the likelihood of severe damage or collapse if such irregularities exist.

5. How is the redundancy factor ρ calculated for walls with h/w < 1.0?

Section 12.3.4.2 states that the redundancy factor ρ must be taken as 1.3 for buildings in SDC D and above unless one or both of two conditions are met. One of these conditions is that "Each story resisting more than 35 percent of the base shear in the direction of interest shall comply with Table 12.3-3." The intent of Table 12.3-3 is that the engineer should consider each lateral load resisting element in each direction and should perform the test associated with that element. For example, consider a system with two moment frames (A and B) and one braced frame (C) resisting loads in a given direction. Analysis would be performed with moment releases placed at each end of a given beam in moment frame A, with frames B and C intact. If the placement of the releases does not reduce the system strength by more than 33 percent or cause an extreme torsional irregularity, the redundancy factor can be taken as 1.0. Theoretically, the test must be performed once for each beam in frame A, then again for each beam in frame B, and then again for each diagonal in the braced frame. Fortunately, the strength check may often be avoided by inspection. However, the extreme torsion irregularity check may not be as easy to visualize.

There is no requirement to remove walls that have a height-to-width ratio of less than 1.0. Thus, if the structure described above had two moment frames and one wall, and the wall had h/w less than 1, the wall would never need to be removed in the redundancy analysis. If the system consisted only of walls, and each wall had h/w less than 1, no walls would need to be removed and the redundancy factor would default to 1.0. This situation would even be the case for a system with only one or two walls (with h/w less than 1) in a given direction. It would seem that such systems are not particularly redundant and should be designed with $\rho = 1.3$. It appears that this system has fallen through the cracks in the ASCE 7 resulting in a potentially unconservative design.

6. Are P-delta effects calculated based on the initial elastic stiffness, or are they analyzed at the design story drift?

The stability coefficient θ computed by Eq. 12.8-16 has two uses. First, it is used to determine if it is necessary to include P-delta effects in the analysis. If θ is less than or equal to 0.1, P-delta effects may be neglected, and if θ is greater than 0.1, such effects must be included. Secondly, when $\theta > 0.1$, θ is used to amplify both the displacements and the member forces, where the amplification factor is given by $1/(1 - \theta)$.

The stiffness used to calculate displacements in Eq. 12.8-16 should be the same stiffness used to compute the period of vibration of the structure and to compute the design story drifts (used in accordance with the allowable story drift of Table 12.12-1). The analytical model used to compute the stiffness should conform to the requirements of Section 12.7.3. These modeling requirements are consistent with a structure subjected to service level loads.

7. Why in earthquake engineering, is the P-delta stability index θ required to be less than $0.5/(\beta C_d)$? This results in permitted P-delta effects much less than those which are allowed in nonseismic design. Considering that no failures have been reported because of P-delta during earthquake, why should such a strict criterion generally control the design?

The limit on the stability coefficient provides two effects. First, it protects buildings in low seismic hazard regions against the possibility of postearthquake (residual deformation triggered) failure. Second, it provides a limit in the implied overstrength of a building.

Regarding P-delta triggered failures, P-delta effects are generally more critical in buildings in low and moderate hazard areas (where buildings have relatively low lateral stiffness) than they are in high hazard areas (where the stiffness is relatively high). There is limited knowledge on the likely performance of code-compliant buildings in the lower hazard areas, but without some limit on the stability ratio, it is entirely possible that failures may occur because of dynamic instability.

The term β in the stability coefficient is essentially the inverse of the overstrength of the story. When computed, the overstrength often exceeds 2.0. Thus, it may be beneficial to compute the β factor when it appears that the upper limit on θ is controlling. Unfortunately, computing the story overstrength is not straightforward (see FAQ 3).

8. **Is it necessary to check P-delta effects (per Section 12.8.7) when such effects are automatically included in the structural analysis?**

When the P-delta analysis is performed by a computer, the displacements and story shears are automatically amplified, so there is no need to amplify these quantities using the ratio $1/(1 - \theta)$. However, it is necessary to determine if the maximum allowable value of θ, given by Eq. 12.8-17, has been exceeded. To do this, it is necessary to recover the stability coefficient from the structural analysis. This coefficient can be recovered by performing the analysis with and without P-delta effects and computing the interstory drifts from each analysis. If the interstory drift from the analysis with P-delta excluded is designated Δ_o and the interstory drift from the analysis with P-delta effects included is Δ_f, the story stability coefficient is given by

$$\theta = 1 - \frac{\Delta_o}{\Delta_f} \tag{F4}$$

If the computed value of θ for each level is less than 0.1, the analysis may be rerun with P-delta effects turned off. If θ is greater than θ_{max} for any story, the structure must be reproportioned.

9. **How much eccentricity in eccentrically braced frames (EBFs) is required for a C_t value of 0.03 and x = 0.75 for determination of the approximate period?**

Table 12.8-2 provides parameters C_t and x to be used in the determination of the approximate period, T_a. The parameters $C_t = 0.03$ and $x = 0.75$ are applicable to eccentrically braced frames. According to the definition provided in Chapter 11, an eccentrically braced frame is "a diagonally braced frame in which at least one end of each brace frames into a beam a short distance from a beam-column or from another brace." Therefore, there is nothing preventing the designer from using, say, a 6-in. eccentricity, which is obviously not the intent of the provisions and would result in a frame stiffness closer to that of a concentrically braced frame than that of an eccentrically braced frame. A convenient definition to prevent the misuse of these provisions is to limit their use to EBFs detailed in accordance with the requirements of the *Seismic Provision for Structural Steel Buildings* (AISC 2005a).

Note that a frame with eccentric connections could technically be used as part of an R = 3 "steel system not specifically detailed for seismic resistance" (System Type H in Table 12.2-1). However, in this case the designer should pay careful attention to the proportioning of the system so that it follows that recommended for EBFs in the steel *Seismic Provisions*. Additionally, for these systems the ratio of the link length e should be approximately equal to 0.2 times the full beam length L to allow for the intended flexibility (see also Fig. 3-25 of AISC 2005a).

10. What is the approximate period of a dual system?

Table 12.8-2 does not include dual systems (or other combined systems), so such systems automatically default to "All other structural systems," wherein the values for parameters C_t and x are 0.02 and 0.75, respectively. It would seem that these parameters would produce a somewhat low period for a dual system, but given that the stiffness of such systems is likely dominated by the stiffer component (e.g., the shear wall in a frame-wall system), the degree of conservatism is probably not excessive.

Also of concern is the determination of periods for combined systems, such as those shown as Buildings C and D in **Fig. G8-2** in Example 8. In such situations, the default parameters may be excessively conservative, but perhaps this use is warranted as a penalty for using nontraditional (and not well understood) structural systems.

If the periods for dual systems and combined systems are determined analytically (using a computer), the degree of conservatism that is associated with the use of the default parameters may be reduced by using the upper limit period, $C_u T_a$, when appropriate.

11. Should drifts be calculated at the center of mass or at the diaphragm corners for comparison to drift limits?

According to Section 12.8.6, drifts are defined as the difference between the displacements at the centers of mass of adjacent stories. However, if the structure is assigned to Seismic Design Category C or higher and has a torsion irregularity or an extreme torsion irregularity, Section 12.12.1 requires that the drifts be computed as the difference between displacements at the edge of the story.

When displacements are allowed to be determined at the centers of mass, a problem occurs when the centers of mass do not align vertically. In such cases it is reasonable to define the drift as the difference in displacements at the center of mass of one story, and the displacement which occurs at the vertical projection of that point on the diaphragm of the story below.

It is also noted that Section 12.2.1 states that "for structures with significant torsional deflections the maximum drift should include torsional effects." Any structure that has a torsional irregularity would classify as "having significant torsional deflections" and, hence, inherent torsion. Accidental torsion (with amplification, if necessary) should be included in the analysis used to compute drift.

12. Table 12.2-1 provides design coefficients for cantilever column systems. The various coefficients (i.e., R, Ω_o, C_d, and height limit) depend on the type of detailing used in the cantilever column system. All systems under this section are frames (which are made up of

columns and beams). How can one apply frame detailing requirements to cantilever columns?

The various material specifications provide a number of requirements for the framing systems that are designated in part G of Table 12.2-1. For example, for columns in special reinforced concrete moment frames, ACI 318 (2005) places limits on material properties, cross-sectional dimensions, area of reinforcement, spacing of reinforcement, and location of splices. Additionally, detailing requirements are provided for detailing and spacing of transverse reinforcement. These limitations and requirements would be applicable to cantilever systems with a special reinforced concrete moment frame as the designated seismic force resisting system. Certain rules, such as strong column–weak beam requirements, are clearly not applicable to cantilever systems (which do not have beams).

Section 11.2 defines the cantilevered column system as "A seismic force-resisting system in which lateral forces are resisted entirely by columns acting as cantilevers from the base." Section 12.2.5.2 places limitations on the axial load that can be carried by the cantilever system and requires that the foundation for such systems be designed with the applicable overstrength factor Ω_o (which is 1.25 for all cantilever systems constructed from steel or concrete). Finally, cantilever systems are subject to the redundancy requirements of Section 12.3.4.2.

13. **I am designing a pedestrian bridge support (hammerhead type) where in longitudinal direction, I can have frame action but in the transverse direction, I have to cantilever the support. If the height of the structure is more than 35 ft, no system in Table 12.2-1 under "Cantilevered column systems" would be allowed. Can I designate the support described above as shear wall and use the R, Ω_o, C_d, and height limit provided for shear wall? More importantly, why is there a difference in cantilevered systems and shear wall systems?**

If the support could be designed and detailed in accordance with all of the requirements for a wall, it would seem that a wall system could be used in the transverse direction. It would appear unlikely, however, that an element with a length-to-thickness ratio less than, say, 4 could be detailed to meet all the requirements of a wall. For special concrete moment frames, columns with a length-to-thickness ratio greater than 2.5 (1/0.4) are not allowed and must thereby be designated as walls.

The lower design values and height limitations for cantilever systems occur because of the generally poor performance of these systems. The poor performance (in comparison with walls) is related to low lateral stiffness and a lack of redundancy of cantilever systems.

14. Section 11.8.3, which is applicable in SDC D, E, and F, requires that the geotechnical report include "lateral pressures on basement and retaining walls due to earthquake motions." Does this mean that these pressures can be taken as zero in SDC B and C?

Lateral pressures caused by ground shaking may be neglected in buildings assigned to SDC B and C. Forces and stresses induced by such pressures in buildings assigned to SDC D through F are considered as part of the earthquake load E in the pertinent load combinations of Chapter 2 of ASCE 7.

15. Are there any specific guidelines to meet the provisions of Section 1.4, "General Structural Integrity"?

There are currently no specific requirements for achieving structural integrity in ASCE 7. However, we expect seismic detailing such as that required in Seismic Design Categories C and above to provide continuity, redundancy, and ductility.

The 2009 IBC introduces requirements associated with structural integrity in Section 1614. These requirements are only specified for a limited number of cases, but they could be used voluntarily for any building. It's not certain that the provisions of Section 1.4 of ASCE 7 will be met by following Section 1614, but the overall integrity of the building will be improved.

16. Are there any specific guidelines on separation between buildings in a seismic event? There used to be a guideline in IBC 2000 for buildings in SDC D or greater in which the separation distance was based on the square root of sum of the drifts of the adjacent buildings.

Section 12.12.3 requires that buildings be sufficiently separated to avoid damaging contact under the total deflection δ_x, determined in accordance with Section 12.8.6. This requirement seems to imply that the total separation between two adjacent buildings (1 and 2), to avoid contact, would be

$$\delta_{\text{separation}} = \delta_{x1} + \delta_{x2} \tag{F5}$$

It is important that the deflections in the above equation include the amplification factor C_d. If both buildings were designed to the drift limit, and that limit was 0.02 times the story height (as for "All other structures" in Occupancy Category I or II in Table 12.12-1), the required separation at a given height h would be $0.04h$, which is two times the limit for a single building. For example, for two 10-story buildings with 150-in. story heights, the required separation would be $0.04(10)(150) = 60$ in., or 5.0 ft. One such building could theoretically be placed no closer than 2.5 ft from the lot line.

If the contact between two adjacent slabs is at the same elevation, it could be argued that some contact can be justified because it is unlikely that the resulting pounding will be damaging to the over structural systems, although this point is difficult to prove.

ICC (2006) allowed the separation to be equal to the square root of the sum of squares of the drift for the two buildings:

$$\delta_{Separation} = \sqrt{\delta_{x1}^2 + \delta_{x2}^2} \qquad (F6)$$

This equation accounts for the unlikely event that the two buildings would be moving in exactly the opposite directions, with both buildings reaching their maximum displacement simultaneously.

The 2009 IBC (ICC 2009) adopted the same approach as the previous editions of the *International Building Code*. For the 10-story buildings mentioned earlier, the IBC building separation would be 42.4 in., which is still a substantial gap. For now, the ASCE 7 requirements should be followed for building separation. For jurisdictions that have adopted the 2009 IBC, the square root of the sum of the squares approach may be used. It is anticipated that the IBC rules will be adopted in ASCE 7-10.

17. Is there an inherent contradiction within ASCE 7 in that the effective seismic weight used in computing equivalent lateral forces is based on 25 percent of Live Load (LL) (for storage loads), whereas the gravity effects include at least 0.5LL (up to 1.0LL) for strength design?

First, according to Section 12.7.2, the 25 percent LL requirement is the minimum for storage live loads. It is feasible, if not likely, that higher loads would be used. The reason that a lower LL is used for effective seismic weight than for gravity is that the loads are (probably) not rigidly attached to the structure and would therefore not be subjected to the same lateral accelerations (and inertial forces) as the structure. Additionally, the sliding movement of the storage load dissipates energy, increasing the effective damping in the system. On the other hand, the gravity live load is constant, even when the building shakes laterally. The same situation occurs with partition loads and snow loads, wherein the amount of load included in the effective seismic weight is less than that used for gravity load effects.

18. Why was 65 ft chosen as the height limit in Section 12.2.5.6?

There is no specific reason other than the fact that 65 ft (for five to six stories) is a reasonable delineator between low-rise and high-rise buildings.

19. How does one calculate diaphragm forces (shear, collector, and chord forces) when using modal response spectrum analysis?

Section cuts above and below the floors can be used to determine the diaphragm forces associated with the vertical distribution of seismic forces determined from the MRS analysis. These forces would be applied in a static manner, similar to the forces computed using Eq. 12.10-1. The minimum diaphragm force requirements of Section 12.10 still apply when MRS analysis is used.

A more accurate distribution of diaphragm forces may be achieved from MRS analysis if the diaphragms are physically modeled using shell elements. In this case, the analyst must be able to define a cut through the diaphragm for which net forces (e.g., shear through the cut) are determined for each mode and then combined using SRSS or complete quadratic combination (see Section 12.9.3).

20. How is the overstrength factor Ω_o used? Are elements designed with overstrength factors expected to remain elastic during an earthquake?

The overstrength factor is used in load combinations 5 and 7 in strength design and in load combinations 5, 6, and 7 in allowable stress design (Section 12.4.3.2). This factor is applied only to certain elements, and never to the structure as a whole. Specific cases in ASCE 7 where the overstrength factor is used include:

1. Design of elements supporting discontinuous wall or frames (Section 12.3.3.3). This factor applies only to systems with horizontal irregularity Type 4 of Table 12.3-1, or vertical irregularity Type 4 of Table 12.3-2.
2. Design of collector elements in SDC C, D, E, or F (Section 12.10.2.1).
3. Foundations of cantilever column systems (Section 12.2.5.2).

The various material specifications may also require design with the overstrength factor. For example, Section 8.3 of AISC (2005a) requires that the axial and tensile column strength, in the absence of applied moment, shall be determined on the basis of the amplified seismic load, where the amplified seismic load is defined as that load combination that includes the overstrength factor Ω_o. The overstrength factor must be considered in numerous other cases in steel design. ACI (2005) also refers to the overstrength factor. For example, the factor is used in association with the design of elements of concrete diaphragms (see Section 21.9.5.3 of ACI 2005).

Elements designed with the overstrength factor do not neccesarily remain elastic during an earthquake. However, these elements are expected to suffer less damage than elements not designed with the overstrength factor and have a lower likelihood of failure.

21. What is the purpose of the exponent k in Eq. 12.8-12?

The exponent k accounts, in an approximate manner, for higher mode effects when distributing the design base shear along the height of the structure. A structure with a uniform story height and story mass with $k = 1$ (for relatively short buildings with T less than or equal to 0.5 s) produces a straight-line upper-triangular lateral force pattern, and a structure with $k = 2$ (for relatively tall buildings with T greater than or equal to 2.5 s) produces a parabolic force distribution with increasing slope at higher elevations.

The limit on using the ELF method of analysis for SDC D, E, and F buildings with $T > 3.5\ T_s$ (see Table 12.6-1) is based on calculations by Chopra (2007) that show that the ELF method may be unconservative when $T > 3.5\ T_s$.

22. A new hangar is needed near San Diego Airport to house Airbus A380-size aircraft (W = 235 ft, H = 79 ft) and perform regular cleaning and maintenance jobs (Occupancy Category II). What kind of SFRS can be selected because none of the systems listed in the ASCE 7 Table 12.2-1 seems adequate? Steel metal building system OMF and IMF (both previously allowed under UBC) can't meet the ICC (2006) and ASCE 7 height limit of 65 ft. On the other hand, a steel SMF is not an option at a 250-ft span because of width-to-thickness limits stated in the AISC 341-05 Table 8.1 (it appears that SMF was formulated for typical conventional construction rather than large clear span moment frames). Braced frames, truss moment frames, or any similar listed system falls short of this job requirement because of height of constructability concerns. What SFRS should we use?

The described structural system is quite special and falls outside the basic intent of the building systems described in Table 12.2-1. However, Section 12.2 states that "Seismic force-resisting systems that are not contained in Table 12.2-1 are permitted if analytical and test data are submitted that establish the dynamic characteristics and demonstrate the lateral force resistance and energy dissipation capacity to be equivalent to the structural systems listed in Table 12.2-1 for equivalent response modification coefficient, R, system overstrength coefficient, Ω_o, and deflection amplification factor, C_d, values." The same clause applies to systems that are described in Table 12.2-1 but violate the height limits.

The demonstrative requirements for an intermediate moment frame, for example, could be met through advanced analysis approaches (e.g., nonlinear response history analysis) and through published test data. Several sets of recommendations for analysis of tall buildings (for which the height limit is often violated) have been prepared (SEAONC 2007; LATB-SDC 2008; CTUBH 2008). These guidelines, prepared from the context of performance-based design, could be adapted for the design and analysis of the large aircraft hangar.

23. **An ice cream kiosk was built in the San Francisco Bay area, on a 25 ft × 25 ft plan, where the back wall is a concrete shear wall and the other three walls are open for glass and doors. The three open walls are designed as steel ordinary moment frames, detailed per AISC 341. Based on Section 12.3 of ASCE 7, the roof diaphragm is neither flexible nor rigid, and a torsional irregularity exists. It appears that Section 12.7.3 calls for three-dimensional modeling and analysis; however, the provision seems to be talking about vertical distribution and multistory buildings. I have a hard time explaining to the owner (and getting paid for) anything beyond common two-dimensional analysis. Is this provision really applicable and intended for one-story structures, especially if it is as small as this job?**

Sections 12.2.5.6 and 12.2.5.7 allow, but place strict restrictions on, the use of single-story ordinary moment frames in SDC D and above.

Regarding the main question, it appears that a three-dimensional analysis would be required to obtain a reasonable distribution of forces in the system, and hence, the requirement of Section 12.7.3 should be followed. (Some kind of 3D analysis was already required to establish the fact that a torsional irregularity exists.) It does not seem necessary, however, to model the roof diaphragm as semirigid (using a computer analysis with shell elements, for example). It is recommended that analyses be performed with both flexible and rigid diaphragm assumptions, and the kiosk be designed for the larger forces arising from the two analyses.

Appendix A
Interpolation Functions

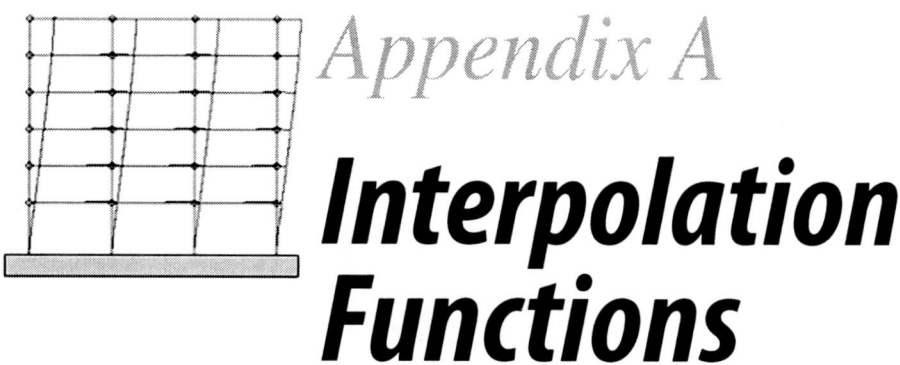

ASCE 7 provides a number of tables from which the user must interpolate to determine the necessary values. This appendix provides a graphical representation of the data provided in several of these tables, as well as mathematical functions from which the appropriate values may be obtained without interpolation.

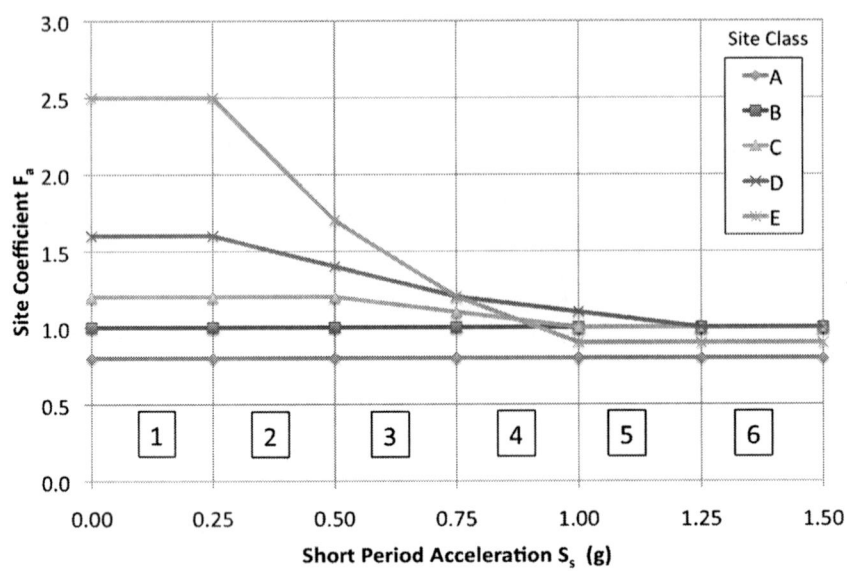

Figure GA–1 Variation of Site Factor Coefficient F_a with S_s.

Table GA–1 Interpolation Formulas for Computing Site Class Factor F_a

Site Class	Value of F_a for a Given Range of Values of S_S					
	(1) < 0.25	*(2)* $0.25–0.50$	*(3)* $0.50–0.75$	*(4)* $0.75–1.0$	*(5)* $1.0–1.25$	*(6)* > 1.25
A	0.8	0.8	0.8	0.8	0.8	0.8
B	1.0	1.0	1.0	1.0	1.0	1.0
C	1.2	1.2	$1.4–0.4\,S_S$	$1.4–0.4\,S_S$	1.0	1.0
D	1.6	$1.8–0.8\,S_S$	$1.8–0.8\,S_S$	$1.5–0.4\,S_S$	$1.5–0.4\,S_S$	1.0
E	2.5	$3.3–3.2\,S_S$	$2.7–2.0\,S_S$	$2.1–1.2\,S_S$	0.9	0.9

Note: See also Table 11.4-1.

Example: $S_s = 0.65g$ and site class C
From **Table GA–1**,

$$F_a = 1.4 - 0.4 S_S = 1.4 - 0.4(0.65) = 1.14$$

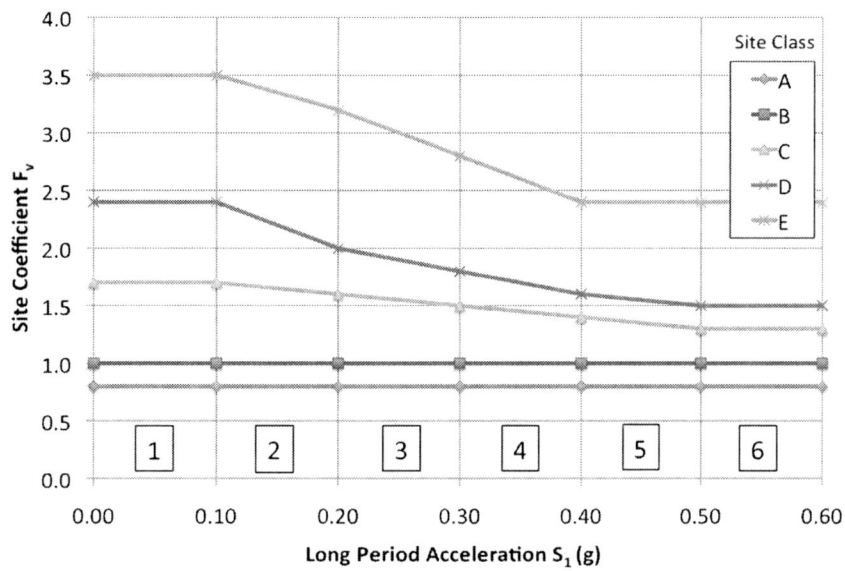

Figure GA-2 Variation of Site Factor Coefficient F_v with S_1.

Table GA-2 Interpolation Formulas for Computing Site Class Factor F_v

Site Class	Value of F_v for a Given Range of Values of S_1					
	(1) < 0.10	(2) 0.10–0.20	(3) 0.20–0.30	(4) 0.30–0.40	(5) 0.40–0.50	(6) > 0.50
A	0.8	0.8	0.8	0.8	0.8	0.8
B	1.0	1.0	1.0	1.0	1.0	1.0
C	1.7	$1.8-S_1$	$1.8-S_1$	$1.8-S_1$	$1.8-S_1$	1.3
D	2.4	$2.8-4S_1$	$2.4-2S_1$	$2.4-2S_1$	$2-S_1$	1.5
E	3.5	$3.8-3S_1$	$4-4S_1$	$4-4S_1$	2.4	2.4

Note: See also Table 11.4-2.

Example: $S_1 = 0.23g$ and site class C
From **Table GA-2**,

$$F_v = 1.8 - S_1 = 1.8 - 0.23 = 1.57$$

Example: $S_{D1} = 0.25g$
From **Fig. GA–3**,

$$C_v = 1.7 - S_{D1} = 1.7 - 0.25 = 1.45$$

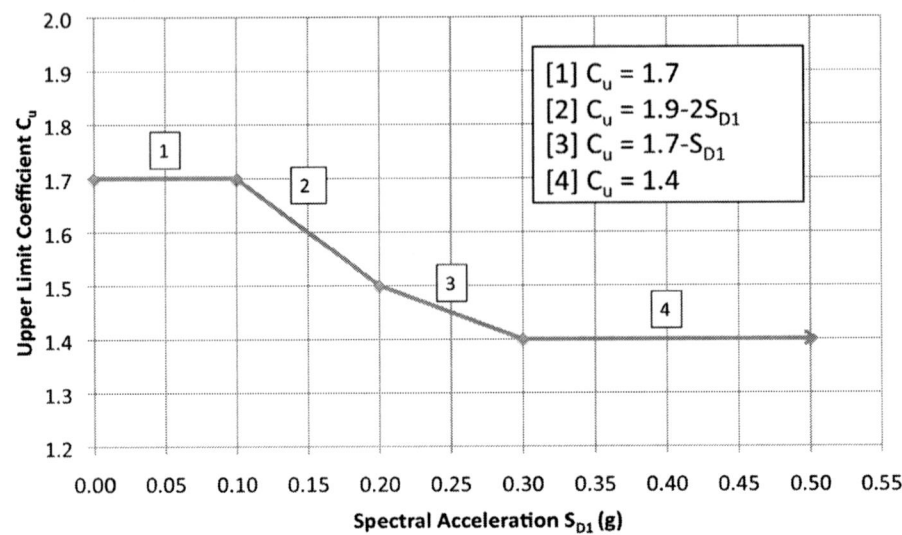

Figure GA–3 Interpolation Functions for C_u (see also Table 12.8-1).

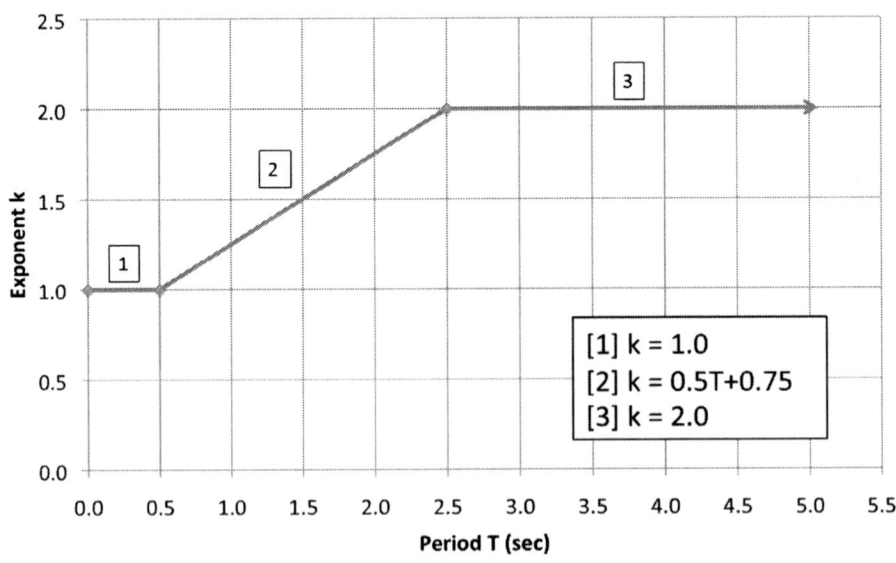

Figure GA–4 Interpolation Functions for Exponent k (see also Section 12.8.3).

Example: $T = 1.7s$
From **Fig. GA–4**,

$$k = 0.5T + 0.75 = 0.5(1.7) + 0.75 = 1.6$$

Appendix B

Using the USGS Seismic Hazards Mapping Utility

The URL for the USGS Hazards Mapping utility is http://earthquake.usgs.gov/research/hazmaps/products_data/. The main page for the USGS Seismic Hazards site is shown in **Fig. GB–1**. A tremendous amount of information is provided through this site, but we will be using only the "Seismic Design Value for Buildings" tools.

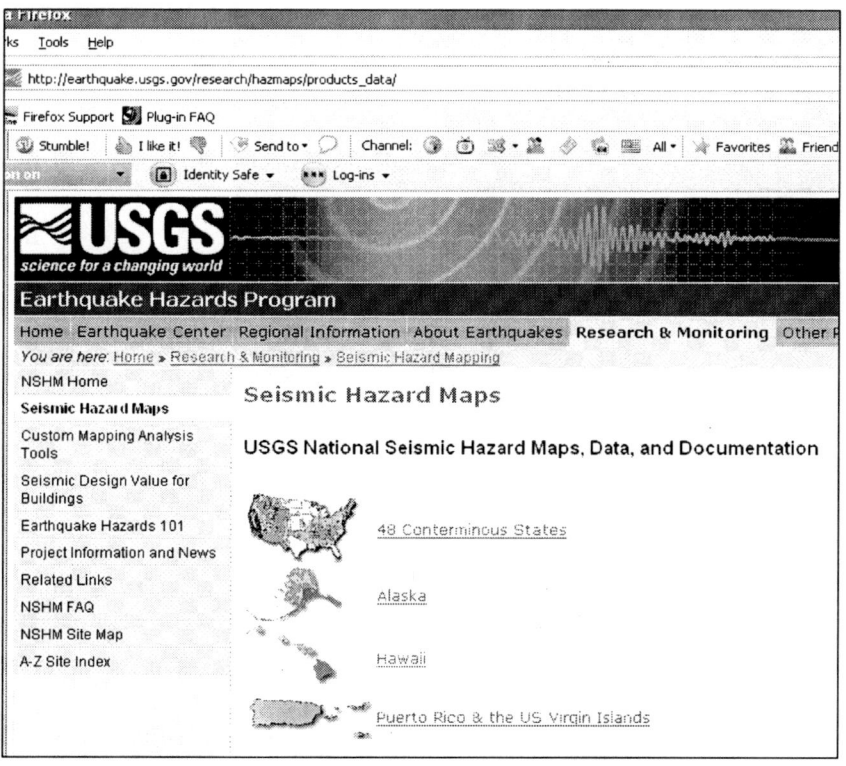

Figure GB–1   Main USGS Earthquake Hazards Page.

This tool provides spectral acceleration values S_s and S_1 for any location in the United States. Values may be given by latitude–longitude, or by ZIP code. We are using Blacksburg, Virginia, as an example.

The Seismic Design Values for Buildings tool is a Java application. The application itself and the Java Runtime Environment must be downloaded and installed on your computer before you can use the application.

To use the tool, click on "Seismic Design Value for Buildings" and the following page will appear:

Figure GB–2 Installation Instructions.

To install Java, follow the instructions shown at the bottom of the page. When you get to the Java page, the application you will need to install is as shown in **Fig. GB–3**.

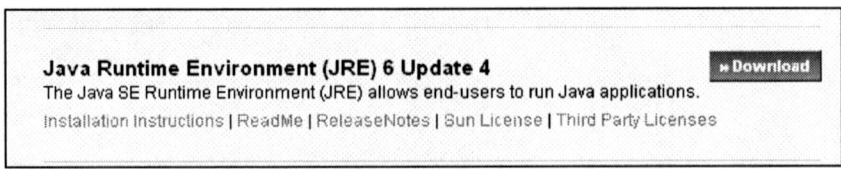

Figure GB–3 Java Environment Download Application.

Before installing the Java Runtime Environment, you will need to uninstall any old versions of this that you may have on your computer.

After installing Java, go back to the Seismic Design Values for Buildings page and click on "Java Ground Motion Parameter Calculator – Version 5.0.8 (4.6 MB)" and the following dialog box will appear:

Figure GB–4 Java Download Application.

Click the "Save to Disk" option, and the application will be installed on your computer. If you prefer, you may run the application directly by clicking on "Open with". You must be online to run this application because it draws information from a USGS server.

If you have done everything successfully, the box shown in **Fig. GB–5** will appear after the utility has been loaded. Click on Okay, and the main program will open. This opening might take a few seconds.

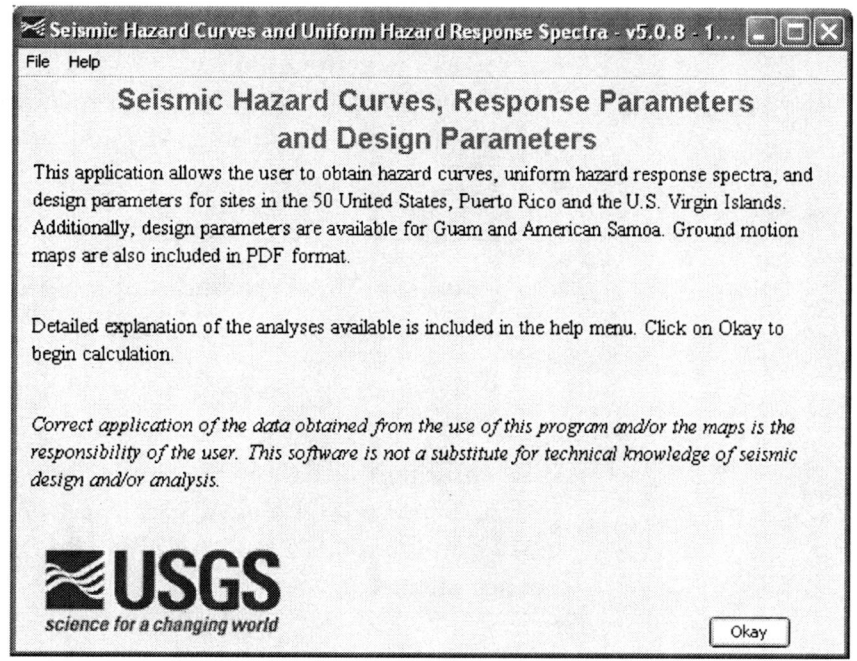

Figure GB–5 Introductory Form.

Seismic Loads: Guide to the Seismic Load Provisions ASCE 7-05 *213*

Now, the screen shown in **Fig. GB–6** should appear. The version shown below the ASCE 7 Standard has been selected under Select Analysis Option at the top of the form (the default in NEHRP). The 2005 version of the Standard in the drop–down list is titled "Data Edition."

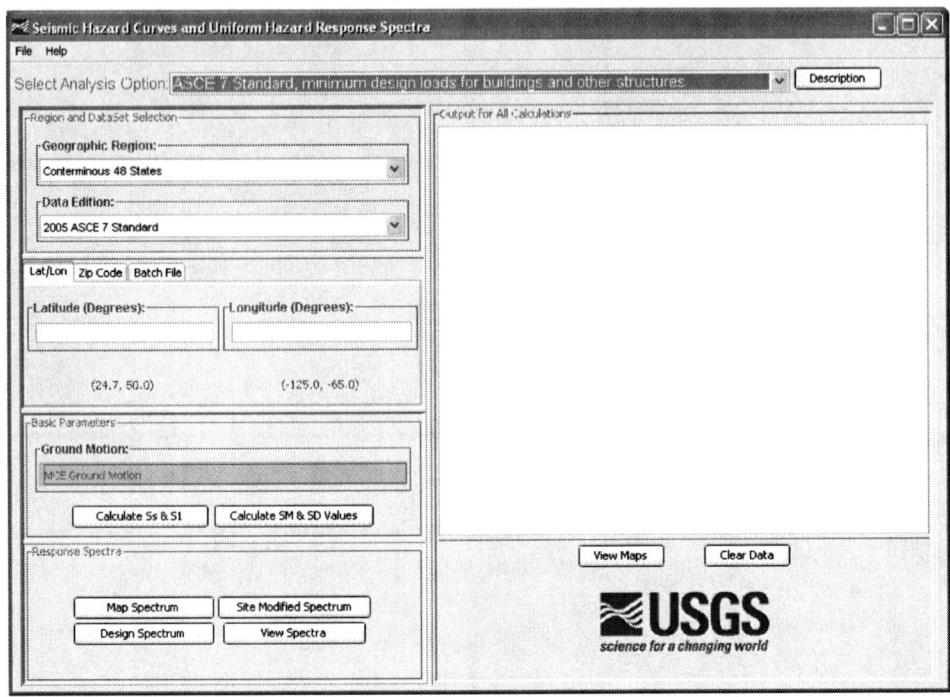

Figure GB–6 Selecting Analysis Option and Data Edition.

The utility has two options for entering a location: ZIP code or latitude–longitude. Caution: when using latitude–longitude, the longitude must be entered as a negative number.[1]

Using *Microsoft Streets and Trips* software (Microsoft 2009), the latitude and longitude for Blacksburg, Virginia, are

Latitude = 37.2
Longitude = –80.4 (negative)

There are Web-based utilities for finding latitude and longitude. For example,

http://stevemorse.org/jcal/latlon.php
http://geocoder.us/

Entering this data in the application and clicking on Calculate S_S & S_1 produces the screen shown in **Fig. GB–7** from which it is seen that S_S is 0.310 g and S_1 is 0.082 g. If the ZIP code (24061) is used instead, similar values are obtained.

---

1. The value must be entered as a negative number because the location is west of the Prime Meridian.

Figure GB-7 Results for Blacksburg, VA.

Finding the ASCE 7 site coefficients F_a and F_v is cumbersome because you usually have to interpolate. Interpolation is easily done using the USGS tool by clicking on Calculate S_M & S_D values. For Site Class E, this action produces the new form shown in **Fig. GB-8**. There are many other options provided by the tool, and it is left to the engineer to explore these options.

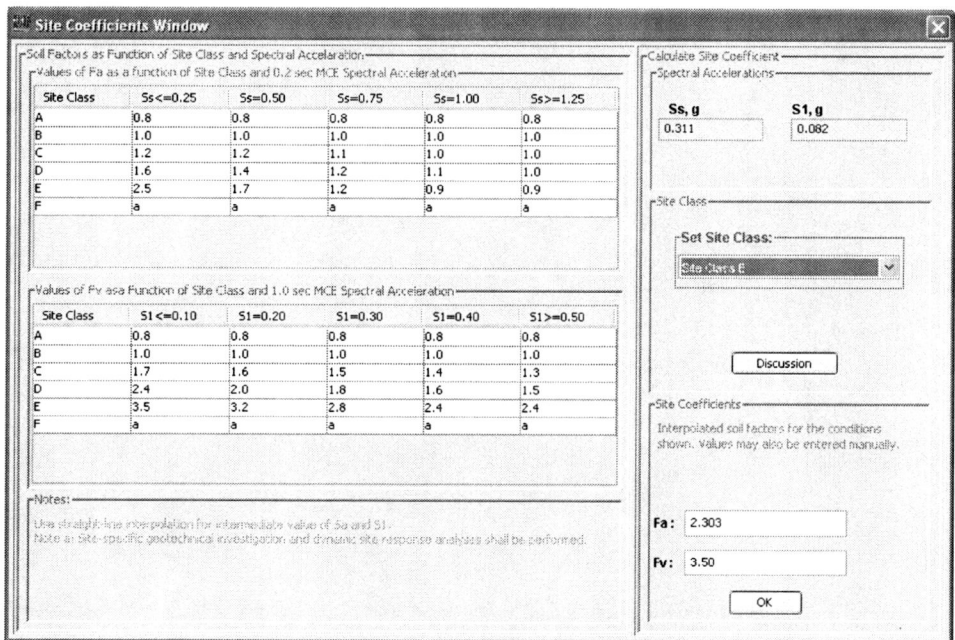

Figure GB-8 Results for Blacksburg, continued.

Appendix C
Using the PEER NGA Database

This document explains how to use the PEER Ground Motion database. In most cases, this database is used to download ground motion record sets to be used in response history analysis of structures.

To access the database, go to the main page for the site, http://peer.berkeley.edu/nga/

To get ground motion records, click on Search (in the middle of the second line of text from top).

To find a specific earthquake, scroll down in the box to the right of "Earthquake." For example, to find the Loma Prieta earthquake, scroll down to "Loma Prieta 1989-10-18 00:05" and select it.

To see where the records are located, click the Search button at the bottom of the form. The results will appear in a map. After zooming in on the map, you will see the map shown in **Fig. GC–1**.

If you click on "Monterey," for example, you will see the map shown in **Fig. GC–2**.

Alternatively, rather than use the map, you may obtain a list all of the ground motion records related to the earthquake. To do this, change "Display Results" from "on Map" to "in Table" and click Search. You will see a long list of record sets, only part of which is shown below:

```
Record      Earthquake                              Station
NGA0731     Loma Prieta 1989-10-18 00:05 (6.93)     CDMG 58373 APEEL 10 - Skyline
NGA0732     Loma Prieta 1989-10-18 00:05 (6.93)     USGS 1002 APEEL 2 - Redwood City
NGA0733     Loma Prieta 1989-10-18 00:05 (6.93)     CDMG 58393 APEEL 2E Hayward Muir
NGA0734     Loma Prieta 1989-10-18 00:05 (6.93)     CDMG 58219 APEEL 3E Hayward CSUH
NGA0735     Loma Prieta 1989-10-18 00:05 (6.93)     CDMG 58378 APEEL 7 - Pulgas
NGA0736     Loma Prieta 1989-10-18 00:05 (6.93)     USGS 1161 APEEL 9 - Crystal Springs Res
NGA0737     Loma Prieta 1989-10-18 00:05 (6.93)     CDMG 57066 Agnews State Hospital
```

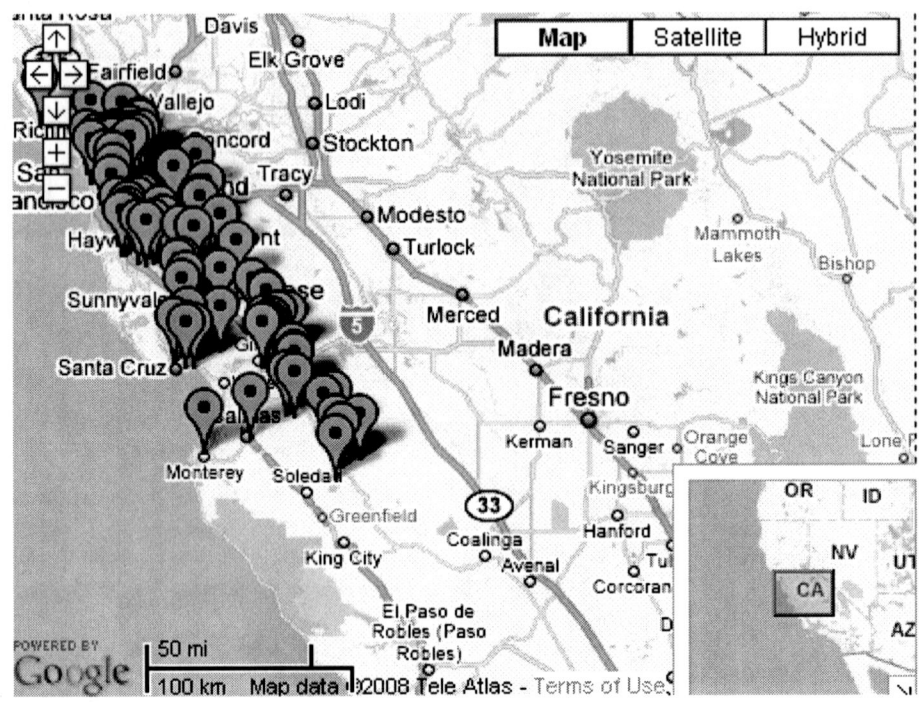

Figure GC–1   Map for Loma–Prieta Earthquake.

Figure GC–2   Selecting the Monterey City Hall Record.

Return to the map-based search display. Using the example search for Monterey, click on "Ground Motion Records" in the bubble caption, and the screen shown in **Fig. GC–3** appears.

This diagram gives information about the earthquake, as well as links to get the actual ground motion acceleration records. There are usually three records per set: north–south (in this case 000), east–west (090), and vertical (UP) records.

To see the north–south ground motion record, click on the record name, for example, LOMAP/MCH000 (displayed in the map-based search results). This is the north–south component because the compass bearing is given as "000" in the title. Compass bearings are shown in **Fig. GC–4**.

The following data will then appear (there are many more lines of data than shown below):

```
PEER NGA STRONG MOTION DATABASE RECORD
LOMA PRIETA 10/18/89 00:05, MONTEREY CITY HALL, 000 (CDMG STATION 47377)
ACCELERATION TIME HISTORY IN UNITS OF G
 7990    0.0050    NPTS, DT
  -0.129203E-02  -0.155257E-02  -0.117803E-02  -0.355531E-03  -0.482559E-03
  -0.484030E-03  -0.481456E-03  -0.477630E-03  -0.476219E-03  -0.479441E-03
  -0.487404E-03  -0.499153E-03  -0.513798E-03  -0.531229E-03  -0.551795E-03
  -0.574890E-03  -0.598821E-03  -0.621665E-03  -0.641146E-03  -0.654379E-03
  -0.658443E-03  -0.651863E-03  -0.635102E-03  -0.610199E-03  -0.580759E-03
  -0.552201E-03  -0.529093E-03  -0.511311E-03  -0.493790E-03  -0.469597E-03
  -0.434634E-03  -0.388477E-03  -0.334624E-03  -0.280486E-03  -0.230926E-03
  -0.187375E-03  -0.152341E-03  -0.129905E-03  -0.122443E-03  -0.129574E-03
  -0.150408E-03  -0.182927E-03  -0.222825E-03  -0.261722E-03  -0.288505E-03
  -0.298566E-03  -0.297028E-03  -0.290274E-03  -0.284506E-03  -0.281662E-03
  -0.276887E-03  -0.266221E-03  -0.251237E-03  -0.242671E-03  -0.249410E-03
  -0.272418E-03  -0.313515E-03  -0.368816E-03  -0.425876E-03  -0.474587E-03
```

The ground motion data is written in rows. In this file, there are 7,990 data points, written at a time increment of 0.005 s. The accelerations are written in g units.

To save the data to your computer, "Select All" of the data, copy, and then paste into a NOTEPAD file.

To use the data in analysis software, you may need to copy the four lines of header information that accompanies each record.

Other information: The PEER NGA database also contains a huge Excel spreadsheet called the "flatfile," which contains a host of information about each ground motion. This spreadsheet can be downloaded from the "Download" link at the top of the file, and the explanation of the data in the spreadsheet can be obtained from the "Documentation" link. This information will be useful when you are trying to select several earthquakes that have similar characteristics, such as fault type, magnitude, site characteristics, and so on.

Earthquake: Loma Prieta 1989-10-18 00:05
Magnitude: 6.93
Mo: 2.7861E+26
Mechanism: 3
Hypocenter Latitude: 37.0407 | Longitude: -121.883 | Depth: 17.5 (km)
Fault Rupture Length: 40.0 (km) | Width: 18.0 (km)
Average Fault Displacement: 108.1 (cm)
Fault Name: San Andreas-Santa Cruz
Slip Rate: 17.00 (mm/yr)

Station: CDMG 47377 Monterey City Hall
Latitude: 36.5970 | Longitude: -121.897
Geomatrix 1: A | Geomatrix 2: G | Geomatrix 3: A
Preferred Vs30: 684.90 (m/s) | Alt Vs30:
Instrument location: BASEMENT

Epicentral Distance: 49.39 (km) | Hypocentral Distance: 52.39 (km) | Joyner-Boore Distance: 39.69 (km)
Campbell R Distance: 44.35 (km) | RMS Distance: 52.51 (km) | Closest Distance: 44.35 (km)
PGA: 0.0700 (g)
PGV: 4.5400 (cm/sec)
PGD: 2.0900 (cm)

ATH	PGA (g)	PGV (cm/s)	PGD (cm)	Filter	nPass	nRoll	HP	LP	Lowest Usable Frequency
LOMAP/MCH000				C	1		0.2	28	0.25
LOMAP/MCH090							0.2	22	0.25
LOMAP/MCH-UP									

Figure GC–3 Detailed Information on Single Ground Motion.

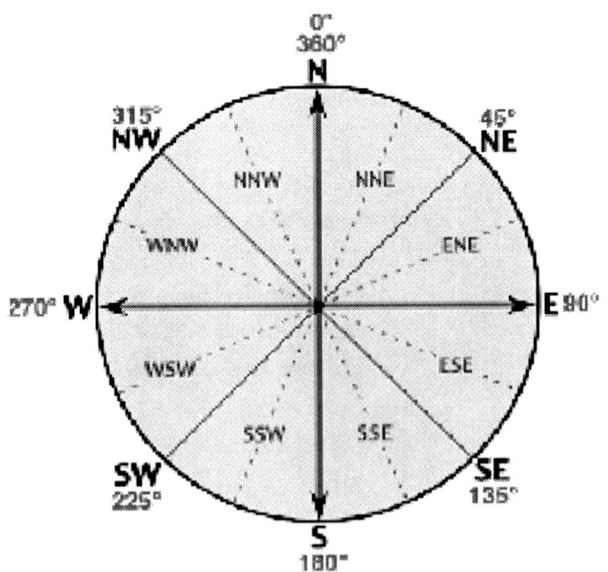

Figure GC–4 Compass for Ground Motion Bearings.

References

American Concrete Institute (ACI). (2004). *Building code requirements for structural concrete*, ACI 318-05 and ACI 318-05R, ACI, Farmington Hills, Mich.

American Institute of Steel Construction (AISC). (2005a). *Seismic provisions for structural steel buildings*, AISC, Chicago.

AISC (2005b). *Seismic design manual*, AISC, Chicago.

AISC (2005c). *Specification for structural steel buildings*, AISC, Chicago.

American Society of Civil Engineers (ASCE). (2006). *Minimum design loads for buildings and other structures*, ASCE/SEI 7-05, ASCE, Reston, Va.

ASTM International (ASTM). (2005a). *Standard test methods for laboratory determination of water (moisture) content of soil and rock by mass*, ASTM D2216-05, ASTM, West Conshohocken, Pa.

ASTM. (2005b). *Standard test methods for liquid limit, plastic limit, and plasticity index of soils*, ASTM D4318-05, ASTM, West Conshohocken, Pa.

ASTM. (2007). *Standard test method for unconsolidated-undrained triaxial compression test on cohesive soils*, ASTM D2850-03a, ASTM, West Conshohocken, Pa.

Carter, C. J. (2009). *Connection and collapse resistance in R = 3 braced frames*, Ph.D. dissertation, Illinois Institute of Technology, Chicago.

Charney, F.A., and Marshall, J. (2006). "A comparison of the Krawinkler and scissors models for including beam–column joint deformations in the analysis of moment-resisting frames." *Eng. J.* (AISC), 43(1), 31–48.

Chopra, A. K. (2007). *Dynamics of structures*, 3rd ed., Prentice Hall, Upper Saddle River, N.J.

Computers and Structures (CSI). (2005). *SAP 2000 theoretical manual*, CSI, Berkeley, Calif.

Council on Tall Buildings and Urban Habitat (CTBUH). (2008). *Recommendations for the seismic design of high rise buildings,* CTBUH, Chicago, Ill.

Federal Emergency Management Agency (FEMA). (2006). *NEHRP recommended provisions: Design examples,* FEMA 451CD, FEMA, Washington, D.C.

Ghosh, S. K., and Dowty, S. (2007). "Bearing wall systems vs. building frame systems." *Go Structural,* <http://www.gostructural.com/article.asp?id=1558> (accessed July 16, 2009).

International Code Council (ICC). (2006). *2006 International building code,* ICC, Country Club Hills, Ill.

ICC. (2009). *2009 International building code,* ICC, Country Club Hills, Ill.

Los Angeles Tall Buildings Structural Design Council (LATBSDC). (2008). *Alternative procedure for seismic design of tall buildings located in Los Angeles,* LATBSDC, Los Angeles.

Liew, J. Y. (2001). "Inelastic analysis of steel frames with composite beams." *J. Struct. Eng.* (ASCE), 127(2), 194–202.

Microsoft. (2009). *Microsoft Streets and Trips,* Microsoft, Redmond, Wash.

Paulay, T., and Priestly, M. J. N. (1992). *Seismic design of reinforced concrete and masonry structures,* Wiley, New York.

Rack Manufacturers Institute (RMI). (2009). *Specifications for the design, testing, and utilization of industrial steel storage racks,* RMI, Charlotte, N.C.

Sabelli, R., Pottebaum, W., and Dean, B. (2009). "Diaphragms for seismic loading." *Structural Engineer,* January.

Schaffhausen, R., and Wegmuller, A. (1977). "Multistory rigid frames with composite girders under gravity and lateral forces." *Eng. J.* (AISC), 2nd Quarter.

Structural Engineers Association of Northern California (SEAOC). (2007). *Recommended administrative bulletin on the seismic design and review of tall buildings using non-prescriptive procedures,* SEAOC, San Francisco.

Wilson, E. L. (2004). *Static and dynamic analysis of structures,* Computers and Structures, Berkeley, Calif.

Index

A

accidental torsion 99–108
    amplification factor (A_x) 103, 105
    diaphragm flexibility 86–87, 106–107, 113
    dynamic loading 108
    equivalent lateral force (ELF) 108, 149
    horizontal structural irregularity 113
    inherent torsion 103
    load combinations 111, 116
    modal response spectrum (MRS) 108, 116, 167
    nodal forces 107
    orthogonal loading 149
    Seismic Design Category 108
    seismic effect, individual member (Q_E) 111
    static analysis 108, 167
    See also torsional amplification
    See also torsional irregularity
approximate fundamental period (T_a) 129–135
    analysis 130
    base 130
    base shear (V) 130, 131, 134
    braced frames 130
    coefficient C_t 130
    combined system 199
    computed period ($T_{computed}$) 132
        v. modified period ($C_u T_a$) 132
    concrete shear wall structures 135
    cracking, effects of 134–135
    deflection amplification factor (C_d) 133
    displacements (δ) 133
    drift 132
    dual system 199
    eccentrically braced frames 198–199
    elastic structures 134
    equivalent lateral force (ELF) 132–133
    exponent x 130
    formula 129–130, 133
    gravity load 134
    ground motions 130
    height (h_n) 130, 131
    lateral forces (F) 133
    lateral load resisitng systems 55–56
    masonry shear wall structures 135
    moment frames 130, 134
        v. braced frames 131
    reinforced concrete buildings 134–135
    seismic analysis 130
    Seismic Design Category 10
    service load stress 134

apr. fundamental period (cont'd)
    shear deformation   135
    shear wall structure   130
    stiffness   133–134
    transition period T_S 130
    use   130
    wind load   130
    X-braced frame   130–131
ASCE 7
    3D analysis   205
    accidental torsion   60, 149
    approximate fundamental period (T_a)   129–130, 135
    base shear (V)   174, 176, 196
    building separation   201, 202
    cantilever column system   200
    computed period ($T_{computed}$)   132, 174
        v. modified period ($C_u T_a$)   132
    diaphragm discontinuity irregularity   62
    diaphragm flexibility   81–82, 114
    diaphragm forces   185, 187–188
    drift   144
    drift computation   155–156
    effective seismic weight (W)   122, 127, 202
    elastic response spectrum   26
    equivalent lateral force (ELF)   70, 202
        two-stage   149
    exponent k 71, 204
    framing systems, different directions   49
    ground motion
        parameters   19
        scaling   30
        selection   30
    height limit   203
    horizontal structural irregularity   57, 67, 114
    importance factor (I)   7, 8
    inertial forces   186–187
    in-plane irregularity   196
    lateral load eccentricity   115
    lateral pressure   201
    load combinations   109–111
        overstrength   110
    load effects   113
    modal properties   171, 182
    modal response spectrum (MRS)   166, 171
    modified period ($C_u T_a$)   132
    moment frames   55
    Occupancy Category   1–2
    one-story structure   205
    orthogonal loading   115–116, 149
    out-of-plane irregularity   196
    overstrength factor (Ω_o)   116, 203
    P-delta effects   158, 163, 198
    period of vibration   129–130
    redundancy factor (ρ)   93–95, 187–188, 196
    reentrant corner irregularity   62
    response history analysis   29
    Seismic Design Category   8, 149, 201
    seismic force-resisting system, special   204–205
    seismic response coefficient (C_s)   140, 141–142, 155–156
    site class   11, 12
    snow load   126
    soft story irregularity   69, 70
    spectral acceleration   19
    stability ratio   158–159, 162, 163
        limit   159
    standard seismic load effect (E)   110–111
    storage live load   126
    story displacement (δ)   174
    story drift   174, 180, 197, 199
    story stiffness   197
    structural analysis   89, 92
    structural integrity   201
    structural systems   41–42, 43
        bearing wall   46
        steel frame   46
        vertical direction   52
    torsional amplification   108
    torsional irregularity   57, 60, 103, 149
    transition period T_S 91
    vertical ground acceleration (E_v)   111
    vertical structural irregularity   67, 69, 79, 190
    weak story irregularity   195
average drift of vertical element (ADVE). *See* diaphragm flexibility

B

base shear (V)
    approximate fundamental period (T_a)   130, 131, 134
    effective seismic weight (W)   143
    equivalent lateral force (ELF)   55, 140, 143
    exponent k 204
    modified period ($C_u T_a$)   174
    period of vibration   55, 155, 174
    redundancy factor (ρ)   94
    response modification coefficient (R)   42, 44
    Seismic Design Category   196
    seismic response coefficient (C_s)   42
    site class   142
    structural systems   44
bearing wall systems   46–48
    building frame systems   47
    response modification coefficient (R)   47

braced frames
   approximate fundamental period (T_a)   130, 130–131, 198–199
   buckling   44
      restrained   130
   coefficient C_t 198
   concentrically   78
   eccentrically   44, 130, 198–199
   equivalent lateral force (ELF)   91
   horizontal structural irregularity   196
   overstrength factor (Ω_o)   196
   seismic response coefficient (C_s)   44
   story strength   78, 161
      overstrength   161
   structural analysis   91
   vertical structural irregularity   196
building separation   201–202

C

cantilever column system   200–201
   design values   201
   frame detailing   200–201
   height limits   201
   overstrength factor (Ω_o)   200, 203
   redundancy factor (ρ)   201
   shear wall system   200
chord forces. *See* diaphragm forces
collector forces. *See* diaphragm forces
combined scale factor (CS_i). *See* ground motion scaling
computed period ($T_{computed}$)
   approximate fundamental period (T_a)   132
   centerline analysis   155
   computing drift   157–158
   modified period ($C_u T_a$)   132, 155
   P-delta effects   158
   period of vibration   174
   story drift   180

D

deflection amplification factor (C_d)   158, 202
design level spectral acceleration   19–22
   Design Basis Earthquake (DBE)   19
   earthquake load effects (E)   110
   elastic response spectrum   26
   example   20–21
   importance factor (I)   142–143
   load combinations   110
   Maximum Considered Earthquake (MCE)   19
   Occupancy Category   142–143
   Seismic Design Category   8
   site coefficient factor   19, 20
   USGS Ground Motion Calculator   22

diaphragm discontinuity irregularity   62–65
diaphragm flexibility   81–87
   accidental torsion   86–87, 106–107, 113
   average drift of vertical element (ADVE)   82
   equivalent lateral force (ELF)   86–87, 107
   horizontal structural irregularity   62–65
   maximum diaphragm deflection (MDD)   82
   modal response spectrum (MRS)   87
   period of vibration   86
   Seismic Design Category   86
   semirigid   86–87
   structural analysis   92
diaphragm forces   185–188
   3D structural analysis   188
   comments   187–188
   inertial forces   186–187
   modal response spectrum (MRS)   188, 203
   redundancy factor (ρ)   187–188
   Seismic Design Category   188
drift effects   153–158
   computing
      computed period ($T_{computed}$)   157–158
      modified period ($C_u T_a$)   155–157
   P-delta effects   159

E

effective seismic weight (W)   119–127
   base shear (V)   143
   building density   125
   dead load   122–125
   equivalent lateral force (ELF)   126, 140, 143, 202
   four-story building   119–127
   industrial building   127
   parameters   122
   Seismic Design Category   143
   seismic load   127
   snow load   126–127
   storage live load   125–126
   vertical structural irregularity   75
elastic response spectrum   25–28
   ground motion scaling   25
   modal response spectrum (MRS)   171–173
   parameters   26
   response history analysis   25
   spectral acceleration   26, 173
   spectral displacement   173
   transition periods   26
   use   25

equivalent lateral force (ELF)   91, 139–151
   3D analysis   126
   acceleration spectrum   140–141
   accidental torsion   108, 149
   approximate fundamental period (T_a)   132, 132–133
   base shear (V)   55, 140, 143
   braced frames   91
   computed period ($T_{computed}$)   144, 174
   diaphragm flexibility   86–87, 107
   displacement, computation of   144
   drift   132, 151
      computation of   144
      effects   153–158
   effective seismic weight (W)   126, 140, 143, 202
   ground motion   141
   inelastic response spectrum   140–141
   lateral forces (F)   141
   lateral load resisting systems, combinations   149–151
   load combinations   113, 114–116
   load effects   113
   mass moment of inertia   126
   modal response spectrum (MRS)   177–180
   moment frames   91
   orthogonal loading   149
   period of vibration   55, 86, 132, 132–133, 144, 149, 150, 204
   redundancy factor (ρ)   149
   reinforced concrete building   144–148
   response modification coefficient (R)   142, 149
   Seismic Design Category   86, 204
   seismic response coefficient (C_s)   140, 141–142
   site class   141
   soft story irregularity   70
   stiffness   149, 150
   story displacement (δ)   174
   structural analysis   89–90
   torsional amplification   149
   torsional irregularity   149
   transition periods   140–141, 142
   two-stage procedure   149–151
   vertical structural irregularity   195
   weak story irregularity   195
exponent k 204
   base shear (V)   204
   interpolation functions   210
   period of vibration   71, 204

F
frames
   See braced frames
   See moment frames
fundamental period scale factor (FPS_i). *See* ground motion scaling

G
ground motion parameters   19–22
   example   20–21
   USGS Ground Motion Calculator   22
   See also spectral acceleration
ground motion scaling   29–39
   2D analysis   31–36
   2D v. 3D analysis   37
   3D analysis   37–39
   combined scale factor (CS_i)   34, 37
   comments   39
   fundamental period scale factor (FPS_i)   34, 36, 37
   ground motion selection   30–31
   PEER NGA database   30, 31
   period of vibration   34, 36
   suite scale factor (SS)   34, 37

H
horizontal structural irregularity   57–67
   accidental torsion   113
   amplification factor   196
   braced frames   196
   consequences   67
   diaphragm discontinuity   62–65
   in-plane offset   196
   load effects   113
   modal response spectrum (MRS)   166
   nonparallel system   66–67
   out-of-plane offset   65
   overstrength factor (Ω_o)   196, 203
   reentrant corner   62
   Seismic Design Category   67
   structural analysis   90, 92
   Type 1a. *See* torsional irregularity
   Type 1b. *See* torsional irregularity
   Type 2   62
   Type 3   62–65
   Type 4   65
   Type 5   66–67
   See also torsional irregularity

I
IBC. *See* International Building Code
importance factor (I)   7–8
   design level spectral acceleration   142–143
   interstory drift   8
   Occupancy Category   7, 8, 142–143
   P-delta effects   158
   seismic strength   8
   use   7
inelastic response spectrum

equivalent lateral force
(ELF) 140–141
structural systems 42
transition periods 140–141
International Building Code 1–2
building separation 202
Occupancy Catagory 1–2
structural integrity 201
interpolation functions 207–210
exponent k 210
site coefficient
long period acceleration 209
short period acceleration 208
upper limit coefficient 210
interstory drift
P-delta effects 158, 198
soft story irregularity 69, 70, 73–75
story stiffness 190
torsional irregularity 103

L

lateral load resisting systems 49–56
approximate fundamental period
(T_a) 55–56
dual system 161, 199
equivalent lateral force
(ELF) 149–151
framing systems, different
directions 49–52
in-plane discontinuity 76
modal response spectrum
(MRS) 55
redundancy factor (ρ) 196
response modification coefficient
(R) 55
stiffness 55–56
story strength 193
structural analysis 55
structural systems, vertical
direction 52–55
torsional irregularity 59
linear response history (LRH). See
structural analysis
load combinations 109–117
accidental torsion 111, 113, 116
amplified seismic 117
design level spectral
acceleration 110
earthquake load effects (E) 110
equivalent lateral force (ELF)
analysis 114–116
horizontal seismic load effect
(E_h) 111
interaction effect 111
lateral load eccentricity 115
live load 110
load effects 113
member forces 116
modal response spectrum
(MRS) 113, 116
orthogonal load 115–116

overstrength factor (Ω_o) 110,
116–117, 203
v. redundancy factor (ρ) 117
redundancy factor (ρ) 111, 117
seismic detailing
requirements 113
seismic effect, individual member
(Q_E) 111
snow load 110
special 110, 116–117
standard 110
v. special 110, 117
standard seismic load effect
(E) 110–111
torsional amplification 111
vertical ground acceleration
(E_v) 111
wind load 110
load types
amplified seismic 117
dead 122–125
gravity 134
lateral 115
live 110, 202
orthogonal 116
partition 202
seismic 127
snow 110, 126, 203
storage live 125–126, 202
vertical 46
wind 110, 130

M

maximum diaphragm deflection
(MDD). See diaphragm flexibility
modal response spectrum
(MRS) 165–182
3D analysis 166, 180–182
accidental torsion 108, 116, 167
degrees of freedom (DOF) 168,
171
design member forces 177
diaphragm flexibility 87
diaphragm forces 188, 203
dynamic analysis 168
effective modal mass 170
elastic member forces 177
elastic response spectrum 25,
171–173
equivalent lateral force
(ELF) 177–180
horizontal structural
irregularity 166
lateral load resisting systems 55
load combinations 116
load effects 113
member forces 116
modal properties 169–171, 182
mode shape 170
orthogonal load 116
period of vibration 55

modal response spectrum (cont'd)
    spectral acceleration   173
    spectral displacement   173
    static analysis   167
    story displacement (δ)   173–174
    story drift   173–174
    story forces   174–176
    story shear   174–176
    structural analysis   90
    structural system modeling   166–169
    torsional amplification   108
    torsional irregularity   166
modified period ($C_u T_a$)
    base shear (V)   174
    computed period ($T_{computed}$)   132, 155
    drift effects   155–157
moment frames
    approximate fundamental period (T_a)   130, 131
    equivalent lateral force (ELF)   91
    Seismic Design Category   55
    seismic response coefficient (C_s)   44
    stiffness   134
    story strength   79, 161
        overstrength   161
    structural analysis   91

N

nonlinear response history (NRH). *See* structural analysis

O

Occupancy Category   1
    ASCE 7 v. IBC   1–2
    design level spectral acceleration   142–143
    exercises   2–6
    importance factor (I)   7, 8, 142–143
    Seismic Design Category   8, 142–143
    story drift   8
    use   1
overstrength factor (Ω_o)   203–204
    amplified seismic load   117
    braced frames   196
    cantilever column system   200, 203
    earthquake load (E)   117
    horizontal structural irregularity   196, 203
    load combinations   110, 116–117, 203
    Seismic Design Category   188, 203
    seismic force-resisting system, special   205
    use   203
    vertical structural irregularity   196

P

P-delta effects   158–163
    2D analysis   162
    computed period ($T_{computed}$)   158
    deflection amplification factor (C_d)   158
    drift   158, 159
    earthquake engineering   197
    importance factor (I)   158
    interstory drift   158, 198
    period of vibration   158
    shear capacity   159
    shear demand   159
    stability ratio   158–159, 162–163, 197, 198
        limit   159, 197, 198
    story drift   162, 197
    story overstrength   159–162, 197
    story shear   162
    story stiffness   197
    total design shear (V_x)   158
PEER NGA database   217–219
    earthquake records   217
    ground motion records   217–219
        scaling   30, 31
period of vibration   129–137
    3D systems   136–137
    acceleration spectrum   140–141
    base shear (V)   55, 155, 174
    centerline analysis   155
    computer calculation   135, 136, 199
    computing with displacements   132–133
    concrete shear wall structures   135
    diaphragm flexibility   86
    drift   132
    eccentrically braced frames   198
    equivalent lateral force (ELF)   55, 86, 132, 132–133, 144, 204
        two-stage   149, 150
    exponent k 71, 204
    fundamental period scale factor (FPS_i)   34, 36
    lateral load resisting systems   55
    masonry shear wall structures   135
    mass participation factor   137
    modal response spectrum (MRS)   55
    modes   136–137
    P-delta effects   158
    response modification coefficient (R)   142
    seismic analysis   130
    Seismic Design Category   10
    seismic response coefficient (C_s)   43–44, 140, 155
    site class   143
    stiffness   133–134, 135
    story displacement (δ)   174
    story drift   174, 197

story stiffness   197
structural analysis   90, 132
structural systems   42
transition periods   142
See also approximate fundamental period (T_a)
See also computed period ($T_{computed}$)
See also modified period ($C_u T_a$)

R

redundancy factor (ρ)   93–98
   1.0 v. 1.3   95
   base shear (V)   94
   cantilever column system   201
   conditions   94
   diaphragm forces   187–188
   elastic analysis   95
   equivalent lateral force (ELF), two-stage   149
   height-to-width ratio   196
   horizontal seismic load effect (E_h)   111
   inelastic analysis   95
   lateral load resisting systems   196
   load combinations   111, 117
   overstrength factor (Ω_o)   117
   Seismic Design Category   93–94, 111, 196
   story shear capacity   195
   system strength   95
response modification coefficient (R)
   base shear (V)   42, 44
   bearing wall systems   47
   equivalent lateral force (ELF)   142
      two-stage   149
   period of vibration   42, 142
   seismic force-resisting system, special   205
   seismic response coefficient (C_s)   42, 44
   site class   142
   structural systems   42
   transition periods   142

S

Seismic Design Category   8–10
   3D analysis   205
   accidental torsion   108, 115
   approximate fundamental period (T_a)   10
   base shear (V)   196
   design level spectral acceleration   8
   diaphragm flexibility   86
   diaphragm forces   188
   effective seismic weight (W)   143
   equivalent lateral force (ELF)   86, 204
   examples   8–10
   horizontal structural irregularity   67
   lateral pressure   201
   load effects   113
   moment frames   55
   Occupancy Category   142–143
   one-story structure   205
   orthogonal loading   149
   overstrength factor (Ω_o)   188, 203
   parameters   8
   period of vibration   10
   redundancy factor (ρ)   93–94, 111, 196
   seismic response coefficient (C_s)   142–143
   site class   8
   story drift   199
   story stiffness   190
   structural analysis   90
   structural integrity   201
   structural systems   42
   torsional amplification   108, 115
   torsional irregularity   60, 113, 115
   vertical structural irregularity   79, 195
   weak story irregularity   195
Seismic Design Value for Buildings tool   211–215
   installing Java   211–213
   interpolating site coefficients   215
seismic force-resisting systems. *See* structural systems
seismic response coefficient (C_s)
   base shear (V)   42
   computing   43–44
   equations   140
   equivalent lateral force (ELF)   140, 141–142
   modified period ($C_u T_a$)   155
   moment frames   44
   period of vibration   43–44, 140
   response modification coefficient (R)   42, 44
   Seismic Design Category   142–143
   structural systems   43–44
shear forces. *See* diaphragm forces
site class   11–18
   base shear (V)   142
   classification   12
   comments   17–18
   determination   13–16
   equivalent lateral force (ELF)   141
   example   16–17
   gathering data   12
   lateral forces (F)   141
   N^- method   15
   N^-_{ch} and $\bar{s}_u$ method   15–16
   period of vibration   143
   response modification coefficient (R)   142
   Seismic Design Category   8
   shear wave velocity   11–12, 14–15
   soil profiles   11, 14
   use   11

site coefficient
    interpolating 208, 209
soft story irregularity 69–75
    drift-based check 70, 73–75
    equivalent lateral force (ELF) 70
    interstory drift 69, 70, 73–75
    stiffness-based check 69, 73–75
    story stiffness 190–193
    weight irregularity 75
spectral acceleration
    elastic response spectrum 26
    mapped (S_1) 42
    Maximum Considered Earthquake (MCE) 19
    structural systems 42
    use 19
    USGS Ground Motion Calculator 22
    See also design level spectral acceleration
stiffness irregularity. *See* soft story irregularity
story capacity. *See* story strength
story displacement (δ)
    computed period ($T_{computed}$) 174
    elastic v. inelastic effects 174
    equivalent lateral force (ELF) 174
    modal response spectrum (MRS) 173–174
    torsional amplification 103
story drift
    calculating 199
    center of mass 199
    computed period ($T_{computed}$) 174, 180
    elastic v. inelastic effects 174
    importance factor (I) 8
    limits 8, 174, 199
    modal response spectrum (MRS) 173–174
    P-delta effects 162, 197, 198
    period of vibration 197
    Seismic Design Category 199
    story stiffness 190
    torsional irregularity 57, 199
    See also interstory drift
story overstrength 159–162
    braced frames 161
    computing 159–162
    dual system 161
    estimating 161
    factors 159
    moment frames 161
    P-delta effects 159–162
    plastic 161
story stiffness 190–193
    comments 193
    interstory drift 190
    P-delta effects 197
    period of vibration 197
    Seismic Design Category 190
    soft story irregularity 190–193
    story drift 190

story shear 190
story strength 159–162
    beam strength 195
    brace strength 193–195
    braced frames 78, 161
    dual system 161
    element capacity 193
    estimating 161
    lateral load resisting systems 193
    lateral strength irregularity 193–195
    moment frames 79, 161
    P-delta effects 197
    plastic 161
    simple systems 193–195
    story shear capacity 195
    vertical structural irregularity 76–79, 193–195
    weak story irregularity 193–195
    See also story overstrength
structural analysis 89–92
    braced frames 91
    computing transition period T_s 90–91
    considerations 92
    diaphragm flexibility 92
    diaphragm forces 188
    equivalent lateral force (ELF) 89–90, 91
    horizontal structural irregularity 90, 92
    lateral load resisting systems 55
    linear response history (LRH) 89, 90
    modal response spectrum (MRS) 89, 90
        v. equivalent lateral force (ELF) 90
    modeling 92
    moment frames 91
    nonlinear response history (NRH) 89, 90
    period of vibration 90, 132
    procedure 89–90
    Seismic Design Category 90
    upper limit period ($C_u T_a$) 90
    vertical structural irregularity 90
structural integrity 201
structural irregularity
    See horizontal structural irregularity
    See vertical structural irregularity
structural systems 41–48
    base shear (V) 42, 44
    bearing wall 46–48
        v. building frame 47
    categories 41
    frames 44
    inelastic response spectrum 42
    parameters 42
    period of vibration 42
    response modification coefficient (R) 42

Seismic Design Category   42
seismic force-resisting system, special   204–205
seismic response coefficient (C_s)   42, 43–44
spectral acceleration   42
steel frame (not detailed for seismic resistance)   46
vertical load   46
suite scale factor (SS). *See* ground motion scaling

T

torsional amplification   99–108
    accidental torsion   103, 105, 115
    amplification factor (A_x)   60–61, 103, 105
    equivalent lateral force (ELF)   149
    load combinations   111
    modal response spectrum (MRS)   108
    Seismic Design Category   108, 115
    seismic effect, individual member (Q_E)   111
    story displacement (δ)   103
    torsional irregularity   115
torsional irregularity   57–61, 103, 105, 115
    amplification factor (A_x)   60–61
    design lateral forces   103
    equivalent lateral force (ELF)   149
    extreme   57, 199
    inherent torsion   103
    interstory drift   103
    lateral load resisting systems   59
    modal response spectrum (MRS)   166
    Seismic Design Category   60, 113, 115
    story drift   57, 199
    See also torsional amplification
    See also accidental torsion
transition periods   90–91, 140–141, 142

U

USGS Ground Motion Calculator   22
USGS Seismic Hazards Mapping Utility   211–215
    installing Java   211–213
    interpolating site coefficients   215

V

vertical structural irregularity   69–79
    amplification factor   196
    braced frames   196
    consequences   79
    effective seismic weight (W)   75
    equivalent lateral force (ELF)   195
    lateral strength   76–79, 193–195
    mass   75
    out-of-plane offset   196
    overstrength factor (Ω_o)   196
    Seismic Design Category   79, 195
    story stiffness   190–193
    story strength   76–79, 193–195
    structural analysis   90
    Type 1a. *See* soft story irregularity
    Type 1b. *See* soft story irregularity
    Type 2   75
    Type 3   75
    Type 4   76
    Type 5a   76–79
    Type 5b   76–79
    vertical geometric   75
    vertical lateral force resisting element   76
    weak story   76–79, 193–195
    weight   75
    See also soft story irregularity